In praise of John Lilly

For the past forty-five years, John C. Lilly has been a peeping Tom at the keyhole of eternity.

As a record of Lilly's search for the nature of reality, *The Scientist* is at times naïve and frustrating, at times profound and inspiring. Despite his fearlessness, curiosity, and perseverance, one senses that Lilly quietly anguishes over his inability to transcend the limits that science has imposed on his quest. Yet his determination is unflagging.

— *New Age Journal*

If there is a cartographer of altered states of consciousness — of the highways and byways of the inner trip — it is John Lilly, a rare combination of scientist and mystic. . . He helps the rest of us break out of the petty orthodoxies that obscure the riches of life from womb to tomb.

— *Psychology Today*

He is erudite, he is driven, he is totally sold on his internal explorations and truths as he agonizes between his scientific self and his inner dimensions and realities. Other writers have made less of a to-do over this dilemma, but others have also been less vital and personal.

— *Library Journal*

BOOKS BY THE AUTHOR

Man and Dolphin (1961)

The Dolphin in History
   (*with M. F. Ashley Montagu*) (1963)

The Mind of the Dolphin: A Nonhuman Intelligence (1967)

Programming and Metaprogramming in the Human Biocomputer (1972)

The Center of the Cyclone (1972)

Lilly on Dolphins: Humans of the Sea (1975)

Simulations of God: The Science of Belief (1976)

The Dyadic Cyclone
   (*with Antonietta Lilly*) (1976)

The Deep Self: Profound Relaxation and the Tank Isolation Technique (1977)

The Scientist: A Novel Autobiography (1978)

The Scientist: A Metaphysical Autobiography (1988)

John C. Lilly, M.D.

SCIENTIST

A Metaphysical Autobiography

Ronin Publishing, Inc. • Box 3436 • Oakland, CA 94609

Published by
Ronin Publishing, Inc.
PO Box 3436
Oakland, CA 94609
http://www.roninpub.com

The Scientist: A Metaphysical Autobiography
ISBN: 0-914171-72-0
Copyright © 1988 and 1997 John C. Lilly, M.D.

Ronin credits:
Project Editors: Sebastian Orfali and Beverly Potter
Cover Design: Brian Groppe
Cover Photo: Philip Bailey
Typography and Production: Ginger Ashworth, Melissa Kirk
Halftones: Norman Mayell

For information regarding rights to works by John C. Lilly write:
InterLicense Ltd., 200 Gate Five Rd., Suite 207, Sausalito, CA 94965.

For information regarding audio and video tapes of John Lilly and
other visionaries, write: Sound Photosynthesis,
P. O. Box 2111, Mill Valley, CA 94942.

The Second Edition was dedicated to Antonietta, whose warmth, love sharing, and diplomacy created a dyadic home, a dyad, a state of being, and an ambiance, encouraging fresh newness in ourselves, in our friends, in our offspring, and in those visitors who resonate with us.

In this Third Edition, I make the following dedications:

To Mary Lilly, for her enduring presence in my life these past sixty years, and creation of our two sons.

To Ani Moss, for her continuing friendship and inspiration.

To Burgess Meredith, for his many years of friendship.

To Elizabeth Lilly, for her many contributions, including the birth of our daughter.

To Rudy Vogt, blessings to my dear friend.

To Michael Kanouff, for a swift and lasting recovery.

And, finally, to Timothy Leary, until we meet again...
dear friend.

John C. Lilly, Maui, Hawaii, 1996

Introduction to the Third Edition of *The Scientist.*

by John C. Lilly, M.D.

Four and one half years ago, after twenty one years in Malibu, California, I moved, with the help of my dear friends, Ann and Jerry Moss, to Maui, in the Hawaiian Islands. Thus, positioning myself to continue my research in a new and exciting location, searching for dolphins and whales in the wild with new friends and old, while continuing to lecture and travel in Europe, Japan, and the United States.

On the sixth of January, 1995, under a clear Hawaiian evening sky, nearly two hundred friends, relatives, Hula dancers, and musicians, gathered to celebrate my eighty trips around the sun. Among them were beloved Mary Lilly, my first wife, with whom I have shared sixty years of love and friendship, along with our two sons, Charles and John Jr.; with John, his beautiful wife and partner of many years, Colette. Cynthia Olivia Roslyn Lilly, my only daughter, attended with her fiance, Wayland "Skip" Augur II.

It was great to see family together, in tropical surroundings, after many years away from one another. John and Colette came from Mexico, where they pursue ethnographic cultural interactions with the Huichol Indians, dedicated to preserving this tribe's traditions. Recently, they presented their research and film archives at the Smithsonian Museum. In 1994, they published a fascinating book of photography documenting the glory of Mexico's Sierra Madre. Charles and his mother Mary came from their ranch in Carbondale, Colorado, where Charles pursues archeological digs and genealogical studies. Mary, I presume, still prays for my redemption, her sense of humor still intact after all these years. Cynthia, named after my friend in the Virgin Islands, Cynthia Major-who was known as the "Angel of Ausschwitz"-continues her MBA studies, and in May of 1995, I proudly attended her wedding to Skip, near their

vii

home in Chico, California.

Also at the gathering were Barbara Clarke Lilly, Nina Carozza Lilly Castellucio, and, in her more ethereal manner, Lisa Lyon Lilly, constituting three of my four adopted daughters. Barbara has travelled at my side through Russia, Japan, Australia, Europe, and the United States on numerous lecture tours, for a variety of causes. She continues her work as a Cetacean Research Scientist, currently creating Ecological World Wide Web Pages, including one dedicated to my research, which can be found at http://www.rain.org/~lili/DrJohnLilly.

She is also the mother of two beautiful daughters, Zoe (currently directing her first movie), and Claudia (an aspiring Marine Biologist). Lisa continues her avid pursuit of knowledge, along with involvement in the arts, entertainment, politics, philosophy, and other infinitems. Nina continues to live at my Malibu home, along with her handsome son, Damon (whom I hope will pursue his many talents). We all miss her mother, Toni, very much. Nina married Anthony Castellucio in July of 1995.

In addition to the family, esteemed friends came from near and far. Dolphin Communication Research Scientist, Louis Herman, of the Kewala Basin Marine Mammal Laboratory, at Waikiki, Oahu, continues important research on visual communication with dolphins for the University of Hawaii. Old friend Oscar Janiger, psychoanalyst and early LSD researcher, came from California with his partner, Kathy Delaney, as did the eminent inventor and metascientist, Arthur M. Young, with his wife, Ruth, and the creativity professor and exemplar, Frank Barron, along with the beautiful Nancy. Tim Leary, who wrote the forward to the second edition of this book, sent his son Zachary as emissary. Tim had just finished his book *Chaos & Cyberculture* (Ronin 1995). Aliya and Craig Inglis, whose wedding I performed in California, were there. Craig is my liason with the Isolation Tank world. He can be reached at 818.876.0218, his email address is eccoscot@aol.com. My liaisons to Japan were in attendance, Gessie Houghton (who has assisted me around the world with his interpretations and inspiring wit (micro-managing and meta-programming), most notably with my continuing media and research projects in Japan, and through two recent sojourns in England), Divyam Preaux (proud owner of a new Isolation Tank in Kyoto (the only one?) and recent companion at Terry Pinney's Dolphins and You "Dolphin Camp" on Oahu), Jean Francois (for transportation and lodgings), and, my Japanese Media Agent, Christopher Langridge, of Media One, Inc., who negotiates my lectures and appearances in Japan and cyberspace. He can be reached via email at lang@shrine.cyber.ad.jp.

The party was organized by one of my best friends, Kutira Decosterd, and my manager, Philip Bailey. Kutira and I, along with her husband, the noted musician, Raphael, search for Whales and Dolphins throughout the year on Maui, and all are welcome to join us, Kutira can be contacted through the

Courtesy of Philip Bailey

At the 5th International Dolphin & Whale Conference, Brussels, Belgium.

Kahua Hawaiian Institute, 808.572.6006, email address k a h u a @ o c e a n i c t a n t r a . c o m , w e b s i t e h t t p : / / www.OceanicTantra.com. Philip has been with me for the last seven years, traveling, arranging, and negotiating my way around the world and my life in paradise. Any questions pertaining to my whereabouts or lecture requests can be channelled through his email address: pbailey@aloha.net.

Since the last update, I've traveled to Australia for the First International Dolphin and Whale Conference, along with Dr. Lawrence "Larry" Raithaus (he, his inspiring and dolphinic wife, Char, and their bright boy, Michael, are my close friends on the neighbor island of Kauai); Dr. Raithaus also translated for me on our trip to Russia, where we attended scientific meetings, including underwater birthing seminars. The conference in Australia was the first of five conferences that have also taken place in America, Japan, and Europe, under the auspices of a vibrant new global organization, "The International Cetacean Education Research Centre," also known as I.C.E.R.C. Founded by my inspiring dear friend, Kamala Hope Campbell, the organization is bridging the gap between human, dolphin, and whale communication, environmental awareness and conservation.

Within the organization, Takako Iwatani, I.C.E.R.C. Japan, Director Of The IVth International Dolphin and Whale Conference, must be singled out for praise for opening the doors to my ideas in Japan, and providing me with many different venues and media opportunities to convey my work. Many thanks go to all of the volunteers of I.C.E.R.C. Japan for their hospitality, friendship and efforts, particularly, Yosinori, Naomi, and Kazuo.

In Europe, Claude Traks, Communicare, I.C.E.R.C. Europe, provided a global audience with a multimedia spectacle for the 5th International Dolphin and Whale Conference in April 1996, at Brussels, Belgium. Barbara Clarke Lilly and I addressed the Conference together for the fourth time. I found our host, Claude, and all of his wonderful volunteer staff of friends at I.C.E.R.C. Europe, as well as the city and people of Brussels, to be delightful, educated, and mindful of our present, past, and future here on Earth as its guardians.

Again, thanks to all I.C.E.R.C. staff globally, the sixth and seventh Conferences take place in Australia, August 1997, and Japan 1998. For more information: I.C.E.R.C. Australia, P.O. Box110, Nambucca Heads, NSW Australia, 2448. I.C.E.R.C. Japan, 3-37-12, Nishihara, Shibuya-ku, Tokyo, Japan. I.C.E.R.C. Europe, rue de Bordeux, 39a, 1060 Brussels, Belgium

My four journeys to Japan for Dolphin and Whale Conferences and other events, have found me lecturing to an enthusiastic, media aware public, who seem to have an energetic vision of applying my ideas within their culture and around the world. I feel that the Japanese can make advances for all of us in interspecies communication as they bring to awareness their massive and belligerent whaling industry, reharnessing their whale harpoons with computer technology to crack the communication barrier between humans and cetaceans. Many of the young people that I have met in Japan have the spark and genuine desire to not only communicate with dolphins and whales, but also live with them in peace.

Five of my books have been translated into Japanese, as well as quite a few other languages. My Japanese book translator, Yasuhiko Suga, has produced a beautiful compact disc of my voice lecturing, with music, entitled, ECCO "The Cogitate Tape", available through Silent Records, 340 Bryant, San Francisco, CA 94107; 415.957.1320. Pioneer Electronics has released an interactive laser disc, entitled "Melon Brains," in which I am included, which features a broad overview of cetacean research and is a fascinating learning tool that teaches the viewer through the eyes numerous research scientists. Sound Photosynthesis, Mill Valley, CA, the work of my good friends, Faustin

Bray, Brian Wallace, and Creon Levit, continues to be a sorcery of available audio and video materials pertaining to my work and ideas, now offering this information over the World Wide Web at http://photosynthesis.com, as well as through catalog, at P.O. Box 2111, Mill Valley, CA 94942.

I have attended the International Whaling Commission meetings for the last three years in Kyoto, Japan, Puerto Vallerta, Mexico, and Dublin, Ireland. At the 1994 meeting, twenty-two nations agreed upon the establishment of a whale sanctuary in the waters around the Antarctic continent. Previously a sanctuary had been established for Sperm whales in the Indian Ocean. Sperm whales, by the way, have the largest brain on the planet: 10,000 grams (I'd still like to know what they're thinking!).

Seventy-one percent of the planet surface is ocean, fifty percent of it is the Pacific, and Maui is in the center of that—making it a grand place for the secretariat of the Cetacean Nation. I am gratified that steps are finally being taken toward recognition of the Cetacean Nation in the United Nations, and that Patricia Forcan, the politically astute President of the Humane Society of the United States, has told me she would agree to be the first Ambassador to it. Every year the Humpback whales migrate through the Pacific, in and around the Hawaiian Islands, spending the winter months here, mating and birthing. Many times I have gone on Whale Adventures in Consciousness with Kutira Decostered and her groups from around the world who swim near the whales. Kutira and her husband, Raphael, appear with a Hawaiian Priest (Kahuna) on their recently released CD "The Calling," mixing music with the songs of the whale along with human voices, among them, my own, which is a very powerful experience.

Further thanks to Adam Trombly and Nancy of Aspen, Colorado (for their work on earthquake prediction and zero state energy) and who have entertained me with their wisdom.

Lee & Glen Perry of Samadhi Tank Company, P.O. Box 2119, Nevada City, CA 95959; phone 916.477.1319, email is samadhi@oro.net. They are responsible for constructing and designing the premier Isolation Tanks available throughout the world now. Kudos to E.J. Gold and friends for publishing our talks on Isolation Tanks in the recent book "Tanks for the Memories," published by Gateways Publishers, P.O. Box 370, Nevada City, CA 95959; phone 916.272.0180.

Gratitude goes out to my literary agent, Manfred Mroczkowski at Interlicense Ltd, Sausalito, CA; email ilicense@aol.com.(for his perseverance), Fond greetings to Mr. Geordie Hormel of Paradise Valley, AZ., fellow Minnesotan, gracious host and friend.

On Maui, thanks to the following friends for helping over the years, as I've adjusted to my new life in the tropics: Shareen Sekota Summers, Lili Townsend and her mother Elcita; Shep Gordon (for his hospitality), Joan Andersen, Daniel Carlson, Brian and Meredith O'Leary, Taylor Fogelquist, Daniel

Courtesy of Philip Bailey

A Meeting of the Minds: John Lilly & Timothy Leary shortly before Dr. Leary's "Deanimation."

McCauley, Adam Keith, Ed Ellsworth at Media Wizards, Milan Param, and Michael Bailey of Greenpeace Hawaii for his help with the "Cetacean Nation."

In Minnesota, my love goes to David and Perrin Lilly, my brother and his wife, lifelong companions, also to David Lilly Jr., and his family. Appreciation to Michael Gilliland, Esq., my counsel there. Duane Feragen deserves gratitude for his steadfast assistance in the financial realms.

In California, to all of my friends at Cybervilla, John Allen, Tango, Rio, Gaye and Lazer (continue with your unique Eco-projects, they are an inspiration towards the possibilities that science must include). Thanks to David Jay Brown for his interview in his wonderful book, "Mavericks of the Mind" and Robert Roth, Esq., my legal counsel in Los Angeles. Across the Atlantic, thanks to my dear friends in Switzerland; Mr. Rudy Vogt (who has shown courage in the face of great opposition, I send my love), Dr. Albert and Anita Hoffman, who recently celebrated his 90th year on the planet! (still a living inspiration to us all), as well as Agnes, Suzanne, and Asti.

Finally, a fond "hello," to my former research associate from the Dolphin Research Project in the Virgin Islands, Margaret Howe, we visited recently for the first time in many years, still searching dolphins and whales; also to Stan Butler and Carol Zahorsky, for their work with Whales Alive, Maui, and the yearly "Celebration of Whales," on Maui; Scott and Hella McVay (friends for many years), and Paul Forestall, Pacific Whale Foundation Maui.

On the evening of May 31st, 1996, my longtime friend, Dr. Timothy Francis Leary, passed into eternity. His spirit was an inspiration to many, his wit, charm, and energy surely will be missed. Farewell, dear friend.

To Sebastian and Beverly at Ronin Publishing, Berkeley, CA, thank you for publishing this new edition of "The Scientist."

In February 1996, Stanford University acquired the John C. Lilly Archive with the help of Robin Rider. It is available for research through Special Collections at Green Library.

On August 20, 1996 the official John C. Lilly World Wide Web site was created by New York artist Bigtwin, Nixy d.b.a. Cybernaut, and the sponsorship of Digital Garage, Tokyo, Japan with the help of Joichi Ito. It can be reached at http://www.garage.co.jp/lilly, and, hopefully by the time of this book's publication at its mirror site at http://www.drjohnclilly.com

So, as I said at the end of the second edition of "The Scientist", "life refuses closure." I hope that my daughter Cynthia will pass on my DNA to a grandchild. And I also hope that there will be greater than ever communication breakthroughs with cetacea to brighten our future. I wish to thank all of my friends, acquaintances, and family for their help, information, and love.

(The substance referred to in the second edition of this book as "Vitamin K" is Ketamine, a short acting psychedelic recommended for use in the isolation tank.)

JCL
August, 1996.
Maui.

With Timothy Leary and Philip Bailey

John Lilly with Timothy Leary, 1994

Foreword to *The Scientist*

by Timothy Leary, PhD

The late twentieth century could well be known as The Second Era of Exploration. An exciting time when human beings uncovered, mapped, navigated, and started civilizing the last frontier: the brain. The birth, you might say, of the species *Homo sapiens, Sapiens*.

John C. Lilly was one of the first scientists to send back systematic and useful data from his explorations of inner geography. For his influential exploits along the uncharted frontiers of knowledge, John C. Lilly will be seen as an heroic philosopher, a veritable Christopher Columbus of the mind.

The First Age of Exploration

To understand the importance of John Lilly's work, it is useful to review the noble exploits of the First Age of Exploration which began at the end of the fifteenth century (1492). For hundreds of years before Columbus, many sages, pilot scientists, people of skeptical commonsense and street smarts, knew that the Earth was round and that land could be reached by sailing west from Europe. Many sailors had made contact with the mysterious western realms, some driven by errant winds, others driven by reckless desires to discover and colonize. There were the Vikings, of course. There was St. Brendan of Ireland, various Celtic sailors, and there was the enigmatic Catalan sea-going mystic, John Ramon Llully.

After 1492 the finest minds and most courageous visionaries of Europe began their series of fabled voyages. Balboa, Magellan, Cortés, Pizarro, de Soto, Champlain, La Salle. These were bold adventurers probing into unknown and scary places, propelled by personal grail goals, greed, piety, power, curiosity, lust for the Fountain of Youth. Each returning to add his notes and observations to construct the map of the New World.

1

And during the later centuries these heroic blazers were followed by millions, stacked in galleys, seeking New Life and New Freedom in the West.

The Second Age of Exploration

Throughout human history the great psychologists and the enduring yogic schools have claimed that the human being possesses, within, a vast world of simultaneous-unifying wisdom, a thought-universe which can be accessed by the skilled adept. These great visionary practitioners developed effective techniques for activating different states of mind, using drugs plus meditative techniques. But they had no understanding that this universe of meaning within was actually the brain.

Until recently there were no models to describe how the brain digitizes, stores, processes, and retrieves information. It was called, "the Mind-Body Problem." Remember?

In the modern west, the first pioneers of the Second Age of Exploration, people like Aleister Crowley, George Gurdjieff, Tom Robbins, Aldous Huxley, Christopher Hyutt, Alan Watts, Baba Ram Dass knew how to use traditional oriental methods for "turning-on" the brain. But they too had no understanding how the brain operated. Indeed, how would it be possible for those living in the mechanical-gears-and-wheels civilization of 1960 to understand how the electronic circuitry of the brain works? Think of the crude neurotransmitters of the 1960s — LSD, psilocybin, ketamine — as the leaky galleons of the first heroic generation of brain pilots.

In these last three decades neurology and psychopharmacology have developed precise methods of activating various altered states with neurotransmitters. The metaphor of brain geography has become the standard paradigm. We speak of left-brain/right-brain. New-brain/old-brain. NLP adepts talk about temporal location, age regression, activat

ing early memories. Researchers are discovering how neurotransmitters switch messages off and on at the synapse.

But Nothing Happens until the Models Are Constructed

In the sixteenth century various national and private
groups sponsored exploration of the New Worlds. They began collecting navigational and cartographic data and attempted to establish colonies. The Hudson Bay Company,
for example. The East India Company.

So it was with the early exploration of the brain. A brave
little band in Menlo Park led by swashbuckling sailors like
Al Hubbard and Willie Harmon launched hundreds of
planned LSD trips, in this case hoping for mental-health prof-
its. In Canada, Humphrey Osmond's lusty lads did the same
hoping for a Nobel Prize. In the film-colony of Hollywood
brave neuronauts like Oscar Janiger M.D. charted the course
of Cary Grant and hundreds of brain-settlers. Around San
Francisco, Ken Kesey and Stewart Brand served as control-
stations for their orgiastic, dionysian Acid Test Flights. The
CIA sponsored a group of dedicated buccaneers hoping to
beat the Russians to the new world. Venerable old Harvard
University became base for the most extensive neurological
missions. Many of these groups developed languages and
models to chart and record psychedelic trips. The Harvard
group alone published over a dozen manuals and guide-
books, based on the *Tibetan Book of the Dead*, the *Tao te Ching*,
etc.

At the same time (1960–1970) over six million landlub-
bers paddled off from the little private islands of their own
minds into the uncharted tropical-arctic realms of the brain.
Each was blown by the winds of set-and-setting to a different continent and returned with a singular report.

All of this feverish activity produced a massive confusion. Something incredibly profound was happening here,
but no one knew exactly what. People were blasting off to

3

bizarre and glorious places, but no one knew where or why.

The "dangers" of brain-voyaging have been hysterically discussed, but not well understood. The perils are not physical. Going to Vietnam in the late 1960s was about ten thousand times more hazardous to your physical health than dropping acid. The suicide rate, which plummeted during the hope-fiend 1960s, predictably rocketed in the sober 1980s.

However. The "dangers" of brain exploration in the early, primitive days of the 1960s were psychological. There were no languages, no models, no explanatory paradigms. At that time the trip involved "terra incognita," mystery, enigma, paradox, perplexity, riddle. The so-called "acid burnout" referred to an acute state of bewilderment, radiation overexposure, short-term memory loss (often complicated by the more insidious "long-term memory gain"). Fortunately this state is always temporary in people with one shred of sanity and common-sense.

The manuals and guides we produced, however scholarly and Hindu-solemn, were woefully inadequate, based as they were on yogic traditions and mystic philosophies that antedated even the industrial age.

During this period of noisy investigation, of high hope and well-publicized confusion there was one man who stood aloof and regally alone. I speak of one scientist who never abandoned the empirical tradition for flashy mystical visions.

This man was Dr. John C. Lilly, M.D. Loner scientist.

In the day-glow psychedelic '60s, you never found J. C. wearing tie-dyes to rock concerts. Nor for him the voyage to the Ganges or to Woodstock. We all saw John C. Lilly as some sort of wizard, a science-fiction starman, a unique back-to-the-future alchemist. A new Paracelsus. A veritable Isaac Newton of the Mind. I am convinced that there has never been anyone quite like J. C. Lilly. He can be understood best in the terms of quantum physics. J. C. is a singularity. A prime number, divided only by himself and One.

4

In 1972 he published a monograph titled *Programming and Metaprogramming in the Human Biocomputer*. This is a work of inestimable importance. Is it too gushy to call this the *Principia Psychologica* of the Cybernetic Age? Few have read it. So what. You don't have to pore over the dusty book. The title says it all. Here, in seven words is the key to the Roaring twentieth century. The updated way to understand the human brain-mind.

Start thinking of your brain as a biocomputer. Wet ware.

Start thinking of your minds (plural) as your software. You know, sloppy disks that you use to process your thoughts and create images on the screens of your consciousness. This is the basic concept of the Cybernetic (Information) Age. I consider it the crowning achievement of post-industrial philosophy. Is not J. C. Lilly the ultimate reality hacker of our era?

Like so many pioneer pilots of the Seven C's J. C. takes off for the frontier, gets lost in the Info-worlds and Cyber-Space matrices. So he doesn't appear for a while. Like don't count on this one to be on time for lunch. But he always arrives "in" time.

His singular realities naturally and righteously reflect his fragile, human origins. The bizarre German-Catholic-alienated loneliness of a boyhood in Biblical background. The frosty, prudish Minnesota. The strict demands of medical training. The Freudian tortures, voluntarily endured. The illuminated intelligence burning like a star.

Above all, we adore the wry, mischievous humanity of the man. *The Scientist* is indeed "A Metaphysical Autobiography." This is one of the great personal stories of our culture, written with the scientific precision that most call honesty. No one has gone as far into the future as John C. Lilly and managed to return (reluctantly, we know) with such clarity and good humor.

Guard your copy of *The Scientist*. It's a precious relic of our wonderful, incredible age.

Timothy Leary
December 1, 1987

Isolation tank, 1996

Contents

Editors' Note

The Scientist weaves hyperspace and history in the narrative. Published in hardcover by Lippincott in 1978, it later appeared as a Bantam New Age Mass Market. This edition incorporates what we call "bubbles" of consensus reality. The bubbles contain photographs and text. They show the personal hard science, and visionary insight, and appear in approximately chronological order.

The editors thank Timothy Leary for his provocative foreword, and acknowledge all those who provided help and material, including Philip Bailey, Barbara Clarke, John Lilly Jr., David Lilly, Charles Lilly, Sandra Katzman, Faustin Bray, Brian Wallace, Laura Huxley, Jim Frohoff, Ed Ellsworth, Richard Feynman, and Burgess Meredith.

Courtesy of Philip Bailey

Dr. Lilly with Barbara Clark-Lilly, Joichi Ito and Kazuo Miyabe, Japan, 1992

John Lilly with dolphin, Oahu, Hawaii, 1994

Acknowledgments

Without the help of his professional colleagues, a few friends, and his best friend, Toni, the author would not have been here to write this book. Thanks to the support and teaching given selflessly by Robert Waelder (Ph.D. in Physics, Vienna), Fritz Perls (M.D.), Richard Price, Burgess Meredith, Grace Stern, Phillip Halecki (Ph.D.), Joseph Hart (ex-S.J.), Will Curtis, Stanislof Grof (M.D.), and several friends who shall remain anonymous, critical times passed safely, delivering the author more or less intact after many transits into other realms of reality. A particular debt of gratitude is owed to professionals who gave consciously directed care to him in times of need: Drs. Robert Mayock, Mike Hayward, Hector Prestera, Craig Enright, Louis J. West, Steven Binnes, and Caroll Carlsen. The professional staffs of five hospitals helped him through painful periods of suffering with standard medical, surgical, and emergency treatment. Esalen Institute at Big Sur, California, extended care, sympathy, and tolerance at critical times. An especial debt of gratitude is expressed to Tobi Sanders, Beatrice Rosenfeld, and Elaine Terranova for their tender loving care in rendering the original manuscript as a more monolithic creation and a more easily read book for the general reader.

The isolation tank was (and is) an essential tool for inspiration, relaxation, and respite from external reality and its multifarious demands/transactions; its definite (though limited) freeing of the body from countergravity forces allowed/allows deep relaxation/sleep for the damaged body and/or the tired mind/brain.

Gratitude is expressed to the dolphins, those pelagic, compassionate, anciently intelligent analogs of Man whose teachings were/are given in wondrously selfless love and tolerance of us. With them we survive; without them, we perish.

Notice to the Reader

From the usual consensus point of view, this book chronicles the inner and outer experiences of the author. Hence it is a continuation of his previous books for those who have read them. Hence it is also autobiographical.

The forms in this work were chosen to free the author for maximum clarity of expression of deeply felt, deeply experienced events. Some of these events involved interactions with powers in the human organizations in which he served. The forms were developed from his necessity to communicate these events without endangering himself and his friends within the human organizations.

He felt, and feels, that the forms used allow expression of a point of view as objectively neutral as is possible to him.

We are all human. No known nonhuman intelligence yet reads what we write or hears what we say. So we are human-centered—communicating only with one another. We judge from human standards. We live for human goals.

We ignore other possible intelligences and communication with them. In a sense, we all unknowingly suffer from a dis-ease which I call "interspecies deprivation." We have no known chroniclers of human events outside our own species, no known judges using nonhuman criteria. We have no history of the planet other than our own. Our economics, our laws, our politics, our sciences, our literature, our organizations are all anthropocentric. Our bodies, in their anthropomorphic form, limit us to human activities, to human communication.

Among the forms of expression used in this work, the author refers to not-human "Beings" whom, in his own deep self, he has experienced at various times. From the human-centered viewpoint of us all, we tend to label such events and their transcription into written form as fiction. In this sense, this book is fiction, an imaginary creation of the author.

Looked at from a less human-centered point of view, all human communication is fiction. We fill in our vast domains of ignorance of the real universe with imaginative explanations. Even our best science was and is created by humans in isolation on 2.9 percent of the dry part of the surface of a small planet in a small solar system in an isolated galaxy.

There are other intelligences—nonhuman—here on the surface of our planet and far more ancient than we are. They are always wet, and we are dry. We kill them for human-centered reasons. Separated from their seaborne communications, we pursue our dis-eased course, deprived of their counsel. We classify them as economic resources to be exploited for human industry. We categorize them as animals, as if we are something or someone else than animals.

Hence, if the author is correct, our current consensus view of whales and dolphins is a contrived fiction, an imagined explanation generated by a sick species—the human one. If the author is incorrect, he is creating a new fiction to be added to all the other works of fiction and is suffering from the overactive imagination to which most humans are heir.

No matter how it turns out in our future, dear reader, please read and enjoy. If you feel safer assuming this book is fiction, do so. If you sense something deeper than fiction, assume it to be so and perhaps you will learn, in a larger perspective, of your own humanness.

Prologue

The Starmaker stirred, awoke from his/her Rest in the Void. Consciousness-without-an-object turned upon Itself, saw Itself turning upon Itself. Feeding back through Itself, it created the *Zeroth Distinction,* an Infinite Series of Itselves alternating with the Void, of the Void protuberating a sequence of Beings$_1$, Void, Beings$_2$, Void, Beings$_3$, Void, et sequens. Out of the Void came HYPERSPACE, the first Sign of the Starmaker awakening Creation.

In the largest sequence, the Starmaker instantaneously (ten to the minus twenty-seventh power seconds) created HYPERSPACE, Consciousness-without-an-object, the *First Distinction* of the Starmaker. HYPERSPACE (Consciousness-without-an-object) is endowed by the Starmaker with his power of Creativity.

HYPERSPACE turned upon Itself simultaneously in two different directions, creating two expanding and contracting vortices within Itself. The Starmaker created the *Second Distinction:* two HYPERSPACE vortices, adjacent, dancing each with the other. There was naught else as yet. The Starmaker, his HYPERSPACE, his two vortices, and the Void were the only Cosmic Contents.

The vortices separated, rejoined, danced together, separated again, rejoined, in infinite sequence. As they joined each other, orgiastic energy built up, closer and closer, higher and higher the positive, loving energy appeared, the *Third Distinc-tion: Love.*

At the peak of the approach of the two vortices they merged in orgasm. Out of their orgasmic meeting spun two new fresh vortices, making now four vortices in HYPERSPACE. The two new ones danced, trading places with the first two, creating new pairs, dyadic creation continuing in HYPERSPACE.

Each pair of vortices was minute (less than ten to the minus thirty-third power centimeters), the beginning of a trace of cer-tainties, of probabilities in a deep sea of INDETERMINACY.

Each pair danced with other pairs as the generative se-quence in HYPERSPACE formed the first Group of pairs. The *Fourth Distinction:* the Group.

Out of the first Group of pairs of vortices in HYPERSPACE came the Origin of space-time, ordinary space, ordinary time. Out of the Origin came the first particle of ordinary matter—a closed group of pairs of dancing vortices, saying "I AM." This group whirled one way, found its mirrored image whirling the other way created samely—the second particle of matter: antimatter. The two groups merged, unwinding their vortices, became the first radiant energy traveling as two photons—two fields fleeing each other, establishing Light in the new space-time.

An infinite sequence of Groups in HYPERSPACE followed, generating a primordial mass of matter at the Origin. As vortical pairs poured out of HYPERSPACE at the Origin, the mass of matter became packed into a Great Bomb, centered at the Origin in space-time. Over a few billions of years of ordinary time the Bomb grew. Suddenly (at a density of ten to the eighteenth power grams per cubic centimeter) it blew up in the first of the series of Big Bangs.

Matter spread, creating more space-time from HYPERSPACE as it spread; it whirled, vorticizing, creating nebulae, spiraling galaxies, stars, planets, all whirling, partaking of the primordial vortices' natures.

The Starmaker watched his Creation, his new expanding creature made of condensed vortices from HYPERSPACE, from Consciousness-without-an-object. He watched as his creature developed Itself, its consciousness of Itself in each and every part.

Communication links, arbitrary signals, appeared, connecting part to part, whole to part, part to whole—consciousness connections between conscious part and conscious whole.

As each small part awoke to its existence in the whole, it assumed it was the center, as if the Origin: that ancient memory built into each vortex pair. Each particle, each atom, each molecule, each organism, each planet, each star, each galaxy thought itself unique, as if the first one of its class. With aging each such Being realized its nonuniqueness, its universality, its creature-nature, its place in the new Universe. Each Being realized its real origin in HYPERSPACE, its constituent vortex dyads.

Each Being existed in two places simultaneously—in ordinary space-time and in a Connection to HYPERSPACE, a Connection to Consciousness-without-an-object, a Connection to the Starmaker. As knowledge of its Self developed, each Being was endowed with a double consciousness, a consciousness of Itself and a consciousness of its Connection. As each Being developed further it remembered the real Origin, the real Connection, and its Universal connected nature. Each Being recognized a portion of its evolution from the past and into the future. Each Being recognized Itself as a portion of smaller wholes and those as a portion of larger wholes.

Within a limited framework of space, space-time, and HYPERSPACE, the Starmaker allowed each to make certain limited choices. The HYPERSPACE portion of each Being remained connected to the whole consciously. Each Being could adjust the amount of Connection to the HYPERSPACE whole and was permitted to forget this for periods of time in which it was involved with the evolution of its own Self. At the end of the evolution of its own Self in a particular form, the Being was to return to HYPERSPACE, rejoin the whole, rejoin Consciousness-without-an-object.

No matter whether the Being was a galaxy, a planet, a star, a solar system, an organism, no matter its unique Self, eventually it became nothing in terms of ordinary matter and space-time as it rejoined Consciousness-without-an-object in HYPERSPACE. It was then permitted to reassemble its individual Self and move

back into ordinary space-time in any Form that it chose, wher-
ever there was room among the other manifested Beings. Some
Beings attaining advanced States of Being were permitted continued conscious memories through all stages from out of HYPER-
SPACE through a current Form and back into HYPERSPACE and out
of HYPERSPACE again.

A Being
Makes a Choice

1

Among the billions of Beings created after the first Big Bang from the primordial Bomb, one took many Forms traveling through the creation of the Starmaker. This Being's portion of Consciousness-without-an-object was segregated from the primordial Consciousness-without-an-object and given the status of an individual consciousness. In its transition from Form to Form, its knowledge was stored in the Network.

Its contained memories were only of how to make transitions and retain memories of its Individuality. In each Form it was permitted to develop new memories based upon its experience. In the infinite sequence of Forms permitted to its choice, at one juncture it chose a particular galaxy, a particular star, a particular solar system, a particular planet, a particular Form on that planet.

It rested in HYPERSPACE, contemplating from HYPERSPACE the planet and the organisms on that planet, and chose its new Form.

On this particular planet were many, many different kinds of organisms, too numerous to count. The Form it chose was that of a human.

It studied its chosen Form, the necessities of conception, of rest *in utero*, its birth to a particular pair of parents. It chose one of two alternate Forms of human, the male or the female.

It chose to be male. From this vantage point it directed a male sperm into a female ovum. It chose the genetic code that would regulate its Form in the future as it grew. It chose the unique genetic code that it would have from those available among the sperms and eggs of the parents.

Once it determined the chain down which it would develop and up which it would go, it moved into a particular sperm and a particular egg and directed the penetration of the sperm into the egg.

As the sperm entered the egg and the nuclear material of each fused into a single cell, there was a vast explosion within the consciousness of this Being. Suddenly it condensed into an individual and proceeded to regulate, to control development— the mitosis and meiosis—of its final Form. It then grew and rested from the task of choice contained within the uterus, allowing the automatic formulation of its new Form by the genetic code.

Birth

The resting Being in the uterus was aware now of its Form: its new Form developed as an embryo and then as a fetus.

Suddenly its new home was limited, limiting. It felt squeezed. It felt compressed. It felt it was dying. The deep redness, purpleness, of its existence was suddenly turned into an unpleasant, devastating constriction. Its local universe became an upheaval, a moving, thrashing, undulating upheaval of its Being. It split off from the local catastrophe, moved out briefly, and watched from outside. It saw clearly the outside of the Mother, saw the Mother struggling to give birth to it. It saw the beginnings of the opening of the birth canal. It saw a head begin to appear. For several hours the head was stuck in the birth canal. Then the Being understood the compression, the devastation, the cutting-off of its Rest. It waited and watched.

Suddenly the head broke through and the body came out. The Being then saw that it was a complete human male baby. It moved back into the body of the baby and activated the respiratory system, gasping the new air into the new lungs as they expanded.

It was inundated with new sensations never before experienced. Explosive amounts of light, of shaking, of shivering, of cold upon a skin which for the first time was felt as an envelope, a boundary, to itself. It heard for the first time its own cries, which at first seemed to be coming from outside. But slowly, surely, it began to realize that by its own voluntary efforts it was producing these cries.

There was a long delay in the cold, in the too-bright lights, in the severance from the warm place.

Suddenly its cries were stifled as its mouth was pushed against a soft, warm surface. Suddenly it started to suck in a warm fluid which restored to it the warm resting-place once again. It felt the warm milk enter into its own Being and become at one with Itself.

Suckling

3

Once filled with milk, it went once more into HYPERSPACE and watched.

From HYPERSPACE it saw a human baby suckling at its Mother's breast, now asleep. It saw the Mother lift the fragile little body and place it in a basket, covering it up to avoid the cold of that particular climate.

It explored the house in which the parents lived and found a small boy playing in another room. In the small boy it saw resentment that the Mother had been taken from him by the new baby. The small boy did not remember his own birth, his own suckling. All he remembered was his weaning, his severance from the Mother at the birth of the new child. His resentment clouded his consciousness and filled him with anger.

The new Being watched as the young boy decided to express his resentment toward the new baby. It watched him go into the room where the baby was in the basket, watched him climb into the basket and begin to pummel the baby. The new Being quickly went back into the baby's body and felt pain for the first time, pain around the head where the boy was beating it. It screamed. The Mother came into the room and removed the small boy, scolding him. The new Being again went out of the body, leaving the sleeping baby.

It watched the little boy get a spanking from the Mother for having mistreated the new baby. The new Being felt the Mother's love for both, her anger at the older one and her protective, enveloping goodwill for the new one. The new Being felt the life-force moving from the Mother in the triadic relationship, fighting for her new dyad, Mother and child.

A long sequence of suckling, of sleep, of terror, of sleep, of suckling went on interminably as the baby grew, as the Being got more firmly attached to the body of the baby. The Being took fewer and fewer trips out of the body of the baby into HYPER-SPACE, maintaining a tenuous connection with HYPERSPACE.

The Being began to listen to the sounds around it. The sounds that the humans made. It stored these experiences of sound; the experiences of sound accumulated to millions and billions.

From its vantage point in HYPERSPACE, it finally realized that the Beings, the human beings, were communicating with words, with sentences, with these peculiar noises that they made. It found that it could mimic these noises in a primitive, long-lasting way with its own mouth; that it could vibrate its own Being with the sounds, not all crying, not all gurglings, the beginnings of vowels, the beginnings of consonants. When in its body, it wondered what all these sounds meant. It watched Father and Mother communicating with one another through these sounds and their bodily expressions.

Slowly but surely these sounds began to have meaning. It began to see that it was called John. That the other male child was called Dick.

The older child was called Dick in a very different way from the "Dick" spoken by its Mother when talking to its Father, who was also called Dick. Gradually the new baby was removed from the basket and put into a crib, where it began to experience its own arms and its own legs, its own body, as if outside itself. It found that if it thought a thought of movement, the movement took place. The body responded to its wishes.

It found it could turn over; it found it could climb on the sides of the crib; it found it could stand up; it found it could see around the room much farther standing than when lying down. It saw the walls of a room for the first time without seeing them from the lying-down position. Once again it found the cold of the night, the darkness. It found the rising sun entering the windows; the light that came and went in a peculiar pattern. It found Time on this planet.

It found that when it felt uncomfortable and cried, its cries were answered by Mother. It found that if it called, Mother came. It found the breasts that Mother gave it to feed it. Other humans were also experienced, Father, Brother, Aunts, Uncles, other children in an infinite sequence.

It found that it could leave the body during the night, and safely; the body went on breathing, its heart went on beating, it went on living. During these periods out of the body, it explored the planet, discovering what minds it could reach. It found its Teachers, the two Guardians. It received instructions about how to behave as a young human growing and learning.

Weaning

4

The young scientist waited in the outer office, considering what he would say on this particular day, what he would analyze of himself with his analyst.

Where did my rage begin? How far back can I go to the beginnings of my rage? How did I decide not to express my rage? When did it become dangerous to do so?

The inner door opened and Robert invited him in. He went into the inner office, lay down on the couch. Robert sat down on the chair at the head of the couch. The young scientist thought a few minutes and finally said. "Today I want to talk about and analyze my weaning experience. I feel that this is important because it was the beginning of my rebellion and of my rage and of my consciousness and of my humanness. I remember a recurrent dream in which I am looking across a room, my mother is in bed, she has an infant at her breast, feeding it. I feel frustrated and enraged, and yet I can't express it. I must hold it in. I am afraid I will be punished if I express it. The baby at Mother's breast is a new baby, just born. I believe that when Mother discovered she was pregnant, she stopped feeding me, weaned me from her breast."

Robert: "How old were you?"

"I was three years old."

Robert: "You mean that you weren't weaned until you were three years old?"

"Yes."

A long silence.

"My parents were Catholic. They did not practice birth control. I found out later that my mother felt if she kept lactating, conception would be prevented."

Another silence.

"I somehow can't speak anymore; the dream is blocked off. I am blocked at the present time. I thought I could talk about this openly and straightforwardly and open up the memories today, but suddenly it is all shut down."

Robert: "What are you feeling?"

"I can't feel anything, all feeling is blocked off. I don't know what to say, my mind is a blank. I am disappointed in myself; I thought I could open up this morning. I can't make it."

Long silence.

"When I was sitting in your outer office I had the feeling that I could get back to the basis of my anger, of my rage at myself, of my guilt, of my fear, of the terror of that three weeks."

Robert: "What three weeks?"

"Of the three weeks of paranoid fear in which everyone was talking about me and in which I was terrified that something or someone would kill me. I stayed in my office and did not allow myself to go out. When I did go out, I felt that they were all against me, were all thinking about me, and were all planning to somehow do me in. I felt this way about my director, about my scientific colleagues, about my wife—anyone and everyone. I realize that this was a totally unreasonable state of mind, that there was nothing I could do about it. I spent most of my time isolated, fearing, crying, and going in and out of this paranoid state."

Robert: "Do you feel there is a connection between that paranoid three weeks and your weaning?"

"Yes, there seems to be a connection, the suppressed rage of the weaning in the dream, the wish to kill."

Robert: "Kill. Kill whom?"

"To kill the baby at her breast. That was my breast, not his.

It was my connection to her. He took my place, she pushed me away—oh, I see, I wanted to kill her too. I did not understand that before."

Silence.

"I ran away. I remember now I watched her feeding him and ran away and hid in a closet and cried and cried. It looks as if I came to a conclusion, a very peculiar conclusion, that I would never again be in a position where I was so needing of someone, something. I denied the need for her breast; I denied the need to replace him with her. I denied the need for her and her love. I cut myself off from feelings for her, for the breast, for the milk, for the attention. I isolated myself from my feelings."

Robert: "I do not like to interrupt, but in order for me to understand this, do you mean that at age three you decided not to feel anything for your mother?"

"That's the way it seems. In the dark closet I made the decision not to have a need for her. The pain and the rage were too great so I made the decision to cut off the feeling, to cut her off at least inside me. Now I suddenly sense that I can't feel positively toward you despite the fact that you are uninvolved and objective. I turn off my feelings or my feelings are turned off by something in me. I see now I do this with everyone. The little boy made the decision, and the man carries it out. It's hard for me to believe this, it's hard for me to have feelings about this. I am afraid that I am going to get angry with you and stop the analysis and leave. Suddenly I am afraid. I am back in the dream watching Mother and the baby. I am shaking with rage and disappointment, totally frustrated. I can't move. I am frozen. I need help. I don't know where to go for help. Where did she go? Where did I go? I am starting to cry; I can barely talk. If I keep this up I am going to break down. But I must keep going; this is a critical point in my development."

Long silence.

"From this point on in my life, the time of the weaning, there is a long blank. I must have been well treated by my parents because I am here today, not locked up, not crazy, so there must have been love there. But from that point on I seem not to have acknowledged it, not to have accepted it, and to have carried on a secret rebellion against love, against closeness, against sharing.

"Later, when I was, I would guess, about six or seven, I was in the church examining my conscience for confession. Suddenly the church disappeared. I saw God on His throne and a chorus of angels singing, and I worshiped Him and felt love for Him on His throne. The two guardian angels were on each side of me taking care of me in the court of our Lord. I feel the awe that I felt then, the reverence that I felt then, and the misplaced love that I felt then. The love of the vision was far greater than love for any human. My lack of love, my lack of feeling, was expressed in love for experiments alone, experiments in electricity and physics and biology, isolated in solitude, alone, experimenting upon matter: frogs, bugs, snakes, plants, transformers, spark coils, chemistry, chemicals.

"Is that why I became a scientist? Was my love for my mother, for my younger brother, so shut off at my weaning that I could no longer share it? Or is this my way of modeling what happened back there? I know the dream is real; that, I can experience firsthand. I know the rage is real, and I know the decision was real.

"I am getting a headache at the site of my migraine attacks."

Robert: "What is your feeling now?"

"I am in my head again, out of my body, and not in contact with my body. All I feel is pain on the right side of my head. I am living in my head, denying the body. I should be feeling rage but I am not; instead there is just pain."

Robert: "When was your last migraine attack?"

"About ten days ago. I thought I wouldn't have another one for eight days. This feels like the beginnings of an attack. But it's out of sequence. It's out of the rhythm of those attacks. So, Doctor, this must be significant.

"I feel that if I could just open up the memories into this region, I could release myself from this suppression of feeling, good and bad feeling, and be free of these migraines. After all, that's why I am here.

"I find my thoughts wandering, going off into considerations of religion, of my Catholic upbringing, of Jungian archetypes, of

Rank's birth-trauma hypothesis. My mind moves away from considering myself, I am getting off into others' theories of why my life is the way it is, including Freud's—including yours—although I am not sure they are yours; my idea of yours, if you wish.

"Evasions! More evasions! Escaping feeling, escaping the trauma, escaping the idea of trauma! I am fed up with myself. I twist, I turn, I evade, I do not penetrate. My mind has defenses a mile deep."

The young scientist looks at his watch and sees that the hour has ended. He gets up and leaves.

The next day he returns, lies down on the couch, and says, "Last night I had a dream. Is that the proper way to begin an analytic session?"

Robert doesn't answer.

"I see you're behaving as a good analyst should. I am feeling angry with you. I feel deprived. I feel you are not giving me the love that I need, the help that I need to get into the weaning problem, to get into why I decided to shut off all feeling for Mother. The more I think about her weaning me, the more reasonable it seems, the more I get into Mother's head and see what her reasoning was—I understand it. My medical training tells me that she had a belief system, unreal as it was, and that the weaning and the previous continuance of feeding me were necessary in her frame of reference. Suddenly I grieve at the ignorance of my parents, at my ignorance. I am sad. I am sad that you are so ignorant, that I am so ignorant, that we need to do an analysis. I would like to be omniscient, I would like to know everything about myself. I would like to know if the vision in the church was real. The nun who put it down, who told me that 'only saints have visions,' my rage at her, my disappointment—I see that's connected with the weaning dream. She was acting the way my mother acted, she was cutting off the feeling of love and awe and reverence. And yet I feel I must forgive her even as I must forgive my mother. Neither one understood, and I was incapable of explaining.

"The dream last night seems to be significant, rather spectacular. I had a slight fever before I went to sleep and I felt hot and childlike. In the dream I am in a weird landscape which is all the color of feces, there is the smell of feces everywhere. I see a desolate feces-colored scene. There is a hot-dog stand at which they are selling feces. There is a feeling of death, of devastation, everything is dying, there is the smell of death everywhere. I am reminded of a scene that actually happened in which I had a bowel movement in my pajamas. The pajamas had feet in them; it was a cold winter. I was lying alone in bed and felt the warm, delightful sensation of feces coming out of me and warming my buttocks. They went down my legs, were warm, soft, and comforting. I went to sleep. My mother woke me up and found the bowel movement. She was very angry, took my pajamas off, and washed me very harshly in the bathtub. I was five or six years old and felt very small and helpless and quite shocked that such a pleasurable thing should be treated with so much rage.

"My mind runs off into all sorts of explanations for such connections—associations, psychosexual development of young children, analytic theories. There I go again. I evade getting into my own memories and my own dreams. I am back in my head. I feel very tense, very reluctant, and very ashamed."

Robert: "Why are you ashamed?"

"I am ashamed of the way I was then, ashamed or rather shamed by Mother, who embarrassed me, who discovered my pleasure in feces.

"I see now that the dream expresses that, doesn't it? Expresses the shame and the fear that the whole universe would become this excrement, the sickness of the excrement. Why do we make excrement so important to our kids? The ideals of cleanliness, personal hygiene, prevention of disease, it's amazing.

"I remember later I put a razor blade into a bar of soap and left it in the bathroom. Mother used the bar of soap and cut her finger rather badly. This caused quite a scene. Mother was very puzzled and I got a bad spanking. Somehow this connects to the dream of the weaning and the episode of being cleaned of my own feces. A hidden revenge. When I try to remember the feelings I had when I was putting the razor blade into the soap, I can't remember them. I was about six or seven at the time. Resentment, vengeance, the sources of same, all seem hidden somehow. Why do I construct all sorts of ideas to cover up what really happened?

"Today I have a feeling that I am one hundred billion cells in search of a leader. I am totally disorganized. Every cell in my body is related in an inexplicable way to every other cell; they seem to be generating a voice which speaks in your presence. Am I merely the voice of a hundred billion cells trying to survive as a unit? Or am I a spiritual being that came to inhabit this vehicle of a hundred billion cells? If I am just a hundred billion cells that originated from the meeting of an egg and a sperm in my mother's womb, which became a fetus, an embryo, an infant, a child, a man, then I am contained within this body. I am the result of the activities of these cells, nothing more and nothing less, with lots of memories which have very little to do with the organization of the vehicle.

"If I am a spiritual being inhabiting the hundred-billion-cell vehicle, then what is this analysis? If I am contained in this body, in this brain, if everything that has happened to me has merely modified and left a trace of past experience within these cells, then the analysis makes sense. The hundred billion cells are trying to become integrated. But if I am a spiritual being inhabiting this vehicle, what is the point of this analysis? The Catholic Church preaches that we have a soul which is separate from the body and which can go on to life everlasting or torture everlasting in Hell— so in a sense the church preaches that analysis is redundant. The soul is a very seductive concept for me. I can use the soul to escape the responsibility for this analysis. I can hear your answer: If the soul does exist, then your choice is Heaven or Hell, based on the way you analyze what's happened to you in the past. Maybe the Catholic Church ought to take up analysis, not the type they used in the Inquisition but the modern, Freudian type of analysis. The material developed could be used in the final judgment of the soul as it leaves the body, and whether it goes to Heaven or Hell would be based on the results of the analysis!

"My God, what bullshit! I do everything to avoid getting at the root of my basic conflict, the suppression of emotion. I keep getting stuck at the weaning. My private theory, which I will share with you as my analyst, is that the weaning was the turning point at which my rather large cerebral cortex shut off access to feelings for others. It chopped the connection. That's a nice neat out, isn't it? I avoid the responsibility for my analysis and for myself by rather devious methods.

"In the vision in the church I was free of the body, I was free of the hundred billion cells, my feeling was free, my reverence, my awe, the expression of what I had cut off for other humans. The young child at the age of three had put into effect a very special program preventing feeling, controlling feeling, cutting off the primitive sources of energy directed at others. He did not succeed fully but enough for it to count in his subsequent experiences with women, with his children, with his colleagues. He saved his energy in this way for his scientific research and became, in a sense, the primordial beginnings of the scientist."

Robert: "So that is the present state of your analysis of the origins of your scientific work, and your dedication to that work? Let us go on with the analysis tomorrow."

The next day the scientist arrived early, sat in the outer office, and considered the previous findings about his weaning. The analysand ahead of him walked through the office and out the door. The scientist got up, walked into the office at the invitation of the analyst, stood above the analyst sitting in his chair, and shouted at him, "How can anybody that's as fat as you are and insists on smoking cigars, which are bad for his blood pressure, analyze me? There you sit, eating too much, smoking too much, raising your blood pressure, and asking for trouble."

The scientist was in a rage. In the midst of his rage he lay down on the couch and shook. There was a long silence. As the scientist calmed down, the analyst spoke. "Your analyst did not have the opportunity of being analyzed by an analyst as clever as your analyst."

The young scientist burst out laughing, realizing that he had carried over his rage, aroused by the previous sessions, and he said, "It's nice to know that my analyst is confident enough to put himself above his analyst in Vienna. I realize that analysis is interminable and that it lasts the length of one's life, and that

here we are merely initiating a new process of examination in
me. Eventually I will be rid of you, even as you were rid of your
analyst. I realize also that my rage at you is the first time I have
experienced rage since I was eight years old. It is relatively safe
to express, at least verbally, this rage at you, knowing that you
can handle it. I also realize that you have not completed your
self-analysis and that I can hardly expect to complete mine
within my lifetime in the way that my perfectionism demands.

"When I was eight years old I went through another decisive experience. My older brother, the same one that pummeled
me in the crib, was baiting me. It was Christmastime and he was
boasting that his presents for Christmas were better than mine. I
had received a small cannon of the type which is fired by adding
water to calcium carbide and pushing on the detonator, causing
a spark that ignites the mixture of air and acetylene to make a
very loud bang.

"My rage at my brother was so extreme that I literally saw
red. I know what that means: one does see red in front of one's
eyes. And I threw the cannon at him so forcefully that, if I had
hit him, I could have killed him. Luckily, the cannon missed his
head by about one inch. Instantly, something took over in me
and said, in effect, You shall never be that angry again, it is too
dangerous. You may kill someone or kill yourself.

"I haven't remembered that episode for the last twenty-six
years and it came to me while I was angry at you. I feel that
there was the same sort of clamping down of feeling, this time of
rage, as there was of feelings of love when I was weaned. Since
then I have felt resentment but never allowed the true, full-
blown syndrome of rage to develop.

"I suddenly feel free to be angry, needing only to control
the expression of that anger in the form of physical damage to
others. There must be whole areas of experience which I have
neglected: feelings of rage, feelings of anger at others which I
have clamped down according to that directive, put into me at
age eight. The directive was too powerful, has exerted too much
influence on my life since then.

"These feelings have been expressed in my life in many different ways: in the church, the vengeful God was projected out there; my guilt when sex started at age twelve was out there; the rage turned around from being inside me to being out there, my fear of Hell represented this suppressed rage. God would damn me to everlasting torture if I gave in to my rage. Rage belonged to God, not to me. I seem to remember a quote from the Bible, 'Vengeance is mine, said God.'

"I said, 'I seem to remember.' That's also a cop-out, isn't it? I catch myself saying to other people, 'I think that,' instead of saying what it is that I think. I defer their harsh reactions by saying, 'I merely think that,' rather than 'I feel that,' whatever it is. I am impressed with the devious nature of the biocomputer which I inhabit. It seems able to escape feeling by all sorts of methods, including language itself and how it is used. I've often wondered why I wasn't more forceful in making demands on people. Now I'm beginning to see some of those programs buried deeply inside which prevent my being forceful.

"It is going to be difficult to integrate this new freedom with all the neglect, over long periods of time, of feelings of rage and feelings of love. Temporarily I expect that I will overact in attempting to get control over this. I will spend too much time trying to release rage inappropriately or/and also to release love inappropriately."

Robert: "We will see how well you do this in your life and we will also see if what you found is significant in your life."

The scientist sat up to leave and looked at the analyst and said, "Do you think it is significant?"

The analyst raised both hands with palms in the air in a gesture of Who knows?

Baby John Lilly

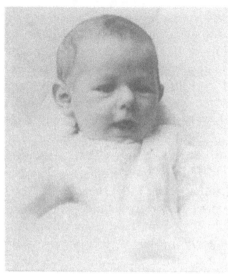

*The baby,
John Cunningham Lilly*

Young toddler.

The three Lilly brothers (from left to right) John, Dick, and David.

John Lilly (right) with his brother David and the family dog Jamey.

courtesy of the Lilly family

John Lilly reaching up to David Lilly held in the arms of their mother Rachael.

Education into Becoming Human

After three years, toward the ending of his psychoanalysis, John walked into Robert's office and said, "I am feeling that I am more independent of you. You have encouraged me to think and feel more and more without your analysis and counsel. My thinking has improved; my feelings are deeper; I am less passive.

"Today I would like to summarize my development insofar as I know it. We have been through much together in this analysis. Somehow I need to integrate it today. I need to live out the important decision points in childhood and through adolescence.

"I'll start with the material we've found in early childhood and work forward in time. The basic conflicts are inner versus outer realities. Here I do not separate these two aspects of my self-struggles. In my deeper inner realities some of these events still live their own lives: I touch them and I am once again within them. Somehow their living being is free of time, imprinted in living memory, molding what I say here."

John, they call me John. Who am I? It's the body they call John, not the Being. The Being comes from somewhere else, inhabits this body. The body is being put through something, something to teach it. What happened last night?

Today I am frightened. Last night I lay in my bed and listened. I heard a scream, Mother screaming, then Father's loud voice trying to calm her. I got out of bed and went to listen at their door. She was moaning and said, "I lost him. I am so tired. I lost another one. What will I do?"

Father said, "We can have another one later. I must bury this one in the backyard."

What were they talking about? Yesterday, Mother's belly was sticking out, today it is flat. There was blood. Another baby?

Mother lost her baby. Father buried it in the backyard. I am frightened.

My little brother is OK. Mother stopped feeding him from her breast last summer. This is spring. She had another baby. Where do the babies come from?

Will Mother make a new baby, or is this the last one?

Dick says Christmas is coming, with presents. It is cold. It snowed last night. I'll put on my snowsuit and go out and make some snowballs.

"Here, Jamey, come on, Jamey, let's go out." Jamey jumps up on me, licks my face, and we go out in the cold snow.

As I approach the wall, suddenly Jamey grabs me by the shoulder and pulls me back. I fall, my shoulder hurts, and I scream with rage at Jamey. That stupid collie dog! He tore my snowsuit. Mother's going to be angry at me. I suppose he thought I was going to fall over the wall. Why can't dogs be smarter? I was just looking over the wall.

"John, how did you tear your snowsuit?"

"Jamey pulled my shoulder. I was leaning over the wall and he bit me in the shoulder and pulled me back."

"Take it off, I have to sew it up for you."

Father wanted to get rid of Jamey and I cried and Mother wouldn't let him get rid of Jamey. Jamey is my best friend.

The Being cut out of the body and looked at the little boy. The little boy saw the Being and said, "Are you my Guardian Angel?"

The Being: "That's what your parents call me, that is not my real name, but it's good enough to use. I am available any time that you need me."

"But I needed you today when Jamey bit me."

"I was there. You were about to fall over the wall and I instructed Jamey to pull you back."

"But I thought Jamey did it because he loved me."

"Jamey loves you but could not understand that you were in such danger. I had to make him do it."

"Will you always take care of me?"

The Being: "Yes, as long as you believe in me. Will you always believe in me?"

"What do you mean 'believe in'?"

"'To believe in' is to know, is to love, is to be with. I am. You are. That is what 'believe in' means."

"I am. You are. I believe in me. I believe in you. Is that what you mean?"

"Yes."

The sun comes into the room, shines in my eyes. I wake up. I hear the birds outside. I cannot go outside. Mommy says I must stay in bed. The nurse will be in in a minute. I can't sit up. I guess she will lift me up again and give me a bath as she did yesterday. They tell me I am sick. I want to go outside and play but they won't let me. Dick and David are outside playing in the yard. I can hear them. I want to go outside....

The nurse came into the room, undressed the little boy, carried him to the bathroom, put him in the bathtub, and washed him. The mother came into the bathroom, looked at the little boy, and started to cry. She said, "Poor little thing, please don't die." (What does she mean, "die"? Does she mean I can't play outside anymore?)

The nurse picked him up, dried him off, and put him back in bed propped up on his pillows. He went to sleep and dreamed.

The Being came and said, "Do you want to go away with me or do you want to stay here?"

"Where will we go?"

"The choice is yours. You can stay here in this body and be the little boy or you can go back with me and join the other Beings."

"Mommy said she didn't want me to die. If I go with you, do I die?"

"That's what dying means, to go with me and leave this place, leave your mommy and daddy, leave Dick and David. Leave Jamey."

"But I don't want to go away. I don't know what you mean, go to the other Beings. I want to get well and play."

"The choice is yours. You will stay here for now and eventually you will go away with me."

"Will you stay with me or are you going away?"

"I will be with you always as long as you believe you will be able to meet with me."

It is summer. The little boy gets up, walks around a bit, gets very tired, goes back to bed. After six months in bed he is allowed to get up and move about. He has gotten to hate the port wine and white of egg that he has been fed. He wants soft-boiled eggs and toast instead.

"Mommy, can I have some eggs?"

"You are getting well! I'll get you some eggs right now."

The powerful electrical storms of summer give way to the colored leaves of fall. School starts.

Kindergarten. First grade. Second grade. Third grade. Miss Ford. Miss Strapp. Miss Curtis. Mrs. Hanke.

Third grade. Miss Curtis. She's beautiful. She loves us. She loves me. Even when I put the blond hair of the little girl in front of me into the inkwell, Miss Curtis understood me and didn't punish me. She just had me wash the ink out of the little girl's hair. Sitting at a desk all day is hard.

There was a bad boy today who won all my agates and my steelies and left me with only clay marbles that won't even roll straight. Dick says agates are best, steelies are next, and clay marbles aren't worth anything. I learned to knuckle down today but I can't hit the marbles out of the ring. I'm going to give up marbles, it's no fun anymore.

Today I went down to the basement and hooked a compass

needle and two electromagnets to a transformer. Whenever I flicked the compass needle on the end of a sharp point, it would whirl and keep whirling. It's like a small electric motor. That's more fun than marbles.

Mr. Dickman at the drugstore is helping me with chemicals. He told me today about sodium and potassium, two metals. If you put sodium in water it melts and travels around on the surface. He gave me some and I tried it in the laundry tub. I put a little bit in. It formed a ball, melted, and traveled around on the surface of the water, making a hissing noise. When I put some starch in the water and a bigger piece of sodium, it got so hot that it burned with a yellow flame and exploded all over the ceiling. Mr. Dickman promised me a piece of potassium.

When I put the potassium on the water in the laundry tub it burned with a beautiful purple flame, kind of violet, fierce, dangerous.

Electricity is powerful. Someone told me that if I used too much electricity I could kill myself. A transformer reduces the amount of electricity so that it is safer. The doorbell on the house works on a transformer. There is an old transformer that I used with the electromagnets from the bell for the compass-needle motor.

Dick got a present for Christmas: When he turns a crank on the generator and someone holds two pieces of metal, one in each hand, it makes one's muscles jump, not very pleasant. He turned the crank too fast for me and I couldn't let go of the metal. That's too much electricity.

Calcium carbide for my cannon. When I put that in water it makes an awful stink, and if I throw a match into the laundry tub the gas that comes from the calcium carbide burns on the top of the water.

Uncle John told me today how the cannon works. The gas coming from the calcium carbide, called acetylene, mixes with air and it explodes and causes the bang when the spark ignites it.

Today I put some calcium carbide in a can with some water and screwed on the top. It exploded. Mother spanked me. She said it was too dangerous to do such things: I could put my eyes out. My Guardian Angel must have been around keeping my eyes away from the can.

Fourth grade. Why did Mommy and Daddy change my school? I don't like this St. Luke's Catholic school. The boys and girls don't play together and the boys play entirely too rough. Today they got me in the middle of a circle, made fun of me, and kicked me.

Fifth grade. Sixth grade. Today I fell out of a tree on my head on the school ground. Mother came and picked me up because she thought I'd been hurt. It was a funny experience.

I was very far away talking to my Guardian Angel. Our conversation was interrupted by somebody moaning and carrying on, crying again and again and again, "Ow, ow, ow." I thought, Whoever that is is interrupting my conversation. My Guardian Angel and I were having a nice talk and this "ow" person interrupted. I slowly began to be in my body and found it was my voice going "Ow." As I came back into my body I opened my eyes and saw that I was at the bottom of the tree.

There is a girl in my class, Margaret Vance, whom I love very much. She has dark brown eyes and dark brown hair. She is beautiful. I watch her all the time. I dream of trading urine with her. I don't know what this means but I have feelings that it would be fun to do this with her.

Father has an exercise machine with a belt on it which shakes his hips when he puts the belt around the back of his seat. When everybody left the house today, I tried the machine. I put the belt around my seat and turned on the motor and it shook my body, and suddenly everything melted and my bottom fell out. The room and the machine disappeared and I went off with my Guardian Angel with my whole body vibrating. There was pain and pleasure, incredible pleasure. My pants got all wet and suddenly I wanted to hide.

Mother came home and found me. She and Father had a private talk and today they took me to the family doctor.

Dr. Brimhall's office. He was the doctor who took care of me when I was sick.

He sat behind his desk, looked me straight in the eyes, and said, "Young man, do you know anything about sex?"

"No." Inside, I thought, He's talking about what happened to me on the machine.

"Your mother and father asked me to tell you about sex. What do you know about masturbation?"

"I don't know anything about masturbation. What is it?"

"At your age boys start to play with themselves and abuse themselves. If they do this too much it can make them insane. Do you do this to yourself?"

"No, I don't know what you're talking about."

I left his office very much ashamed, not knowing the connection between what he was saying and what I had experienced. Mother was waiting outside and I blushed when I saw her.

At St. Luke's they asked for volunteers to be altar boys. I volunteered. I went to the church with Father Smith. He handed me a card with a lot of foreign language on it. He said, "Learn these responses."

"I don't even know how to pronounce the words."

"There is another, older, altar boy here who will teach you how to pronounce them."

With the help of the older boy I learned how to pronounce Latin, not knowing what it meant. Somehow I learned the Mass in Latin. I found a missal which translated the Latin and then I was able to learn it faster.

"Suscipiat, Dominus, sacrificium de manibus tuis ad laudem et gloriam. . . .

"Accept, O Lord, this sacrifice from thy hands for the praise and glory. . . .

"I believe in God, the Father Almighty, Creator of Heaven and Earth. . . . I believe in one, holy, Catholic and apostolic Church."

The Catechism. I believe. The Credo. What I am supposed to believe. I do believe. The Church is the salvation of my soul. I sing in the choir. I behave as an altar boy should. At Mass I put on the proper clothes, I kneel in front of the altar, I answer the proper responses in Latin to the priest at the proper time. I ring the little bell. I worship as I am supposed to worship. I go to confession. I have my First Communion. I go through Confirmation. I remain silent for a day, for twenty-four hours, with my Protestant friends, showing them that I am a saved soul and they are not saved souls.

One day I go to confession at the Cathedral, not at my church, St. Luke's Church.

I enter the confessional, look through the screen at the priest, and start my list of venal sins.

"Today I got angry with Mother because she did not give me my allowance because I had hurt my dog."

I am interrupted by the priest, who says, "Don't you have anything important to report? Do you jackoff?"

Suddenly I am ashamed. I blush. My body feels hot. I am afraid. I don't know what he means and yet something in me knows what he means. I am overwhelmed with my shame. I stumble out of the confessional, walk out of the cathedral, and decide never to go back. If the church believes that it is a sin to have the kind of fun that one can have with oneself I give up the church because it makes no sense whatsoever. My Guardian Angel, please guard me, please, this cannot be wrong.

This summer we moved from the city into the country. We have a farm. We have horses. I have learned to ride a Shetland pony by the name of Dolly. Dolly threw me off and I hurt my head. Mother and Father and Dick were riding with me. They came and picked me up and put me back on Dolly and we rode on home. I hate riding because it hurts my bottom. Yet everybody rides so I must ride.

Dick is building a corral, has bought Western chaps, Western riding boots, and a ten-gallon hat. I prefer to work with electricity, to observe frogs, to learn about anthills. I want to find the snakes in the stream and their nests, and see the little snakes slithering out of a nest when I tear it open. The stars at night are mysterious; the distant galaxies, the northern lights, there is so much more to the world than horseback riding and the incessant chatter of humans.

I met Jack Galt today, his full name is John Randolph Galt.

He has an amateur radio station in his apartment. I am fascinated by the vacuum tubes, by the electricity, and by the fact that with a telegraph key he can talk halfway around the world. He says he will teach me amateur radio.

We work together. I learn about quartz crystals, how to grind a quartz crystal to a given thickness, to create a definite radio frequency. I learn about vacuum-tube circuits. I learn the Morse code. We set up a new radio station in the garage, upstairs. We acquire a new partner, Harry Morton, who works for Northwest Airways.

On the farm that summer I had rabbits. I made a cage for the rabbits which I put on the big lawn and moved around so that they would have new grass to eat.

The newspaper in St. Paul has just been sold to a New York newspaper chain. Father brings the family out that now owns the newspaper. Dick likes Bernard. I hate Bernard. While Dick and Bernard are playing football on the front lawn I fill a Coke bottle with sulfur, sodium chlorate, and charcoal. I drop a match in the Coke bottle. The oxygen formed by the chlorate burning the carbon makes an intense flame which suddenly reaches an explosive velocity. The Coke bottle blows up right in front of me. Not a piece of the glass touches me but a piece goes into Bernard's calf. He screams and grabs his leg.

I am severely lectured by everybody concerned. My shame is great. I disappear for hours into the pasture.

Dick has his own building on the farm, the Granary, it's called. His friends come out and spend the night with him there. I am allowed to sleep there one night. Dick takes his .22 rifle and shoots out the light in the ceiling. He hangs up one of his friends by the thumbs from the ceiling. I run, afraid.

The next day I am out in the pasture wandering around looking at the plants and the animals. I have been digging a cave and I have found some fossils. In the piece of rock that I find, there are many shells and some plants fossilized. I am fascinated. I read the encyclopedia and find that Minnesota was in the glacier region and that these rocks were deposited there by the glacier.

I come out of the cave where I have found these rocks and

suddenly a sound like a very fast bee goes over my head and I hear an explosion. I duck and run back into the cave. I realize that Dick is shooting at me with his .22 rifle from across the pasture. I don't dare come out for the next hour. That evening at dinner I tell Father what happened and Father beats Dick. His rage is overwhelming. The next day Dick beats me up because I told on him. That night Dick goes out in his car, gets drunk, is arrested, and Father beats him again.

Dick gets a new horse, a bronco from Montana which he breaks in the corral using his lariat and the long leather whip which I had helped him to weave. The bronc becomes his horse and he rides it with his friend, Alice, who lives down the road.

One day after riding with Alice, he is late for dinner. We are all sitting down at the table and Father says, "Where is Dick?"

Mother says, "He went riding with Alice today. He's at her house. I called there and he's on his way home."

Father becomes very angry. As time goes on and Dick does not show up, they call Alice's house; he has left there three quarters of an hour ago.

Father leaves, goes across the field looking for Dick, and finds his horse with mud on the saddle at the corral. He goes down across the fields into a valley where there is a swamp and finds Dick there lying in the swamp. Beside him are two deep holes where the bronc's legs went down into the swamp as he galloped off the hillside. An ambulance comes, and Dick is taken to the hospital. He is unconscious and looks dead.

I go down on the lawn to my rabbit cage and cry and cry and cry.

Dick lives three days. The surgeon opens him up but cannot stop the bleeding from his liver. He dies on the table.

I vow to become a doctor and to prevent this from ever happening again to anyone that I love.

I am taken out of the Catholic school and sent to the St. Paul Academy where Dick had been going for the last year. A new life opens up for me.

I take on the bronc, the one-man horse of Dick's. He tries to buck me off and I manage to hold him and stay on. I learn how to control the bronc. I go riding with Alice and fall in love with her. I spend hours with her talking about the universe, about philosophy, about life, about death, about love. I try to make love to her and she gently repulses me, telling me that she is too much older than I am and that I will fall in love with someone my own age.

The summer I am fifteen I fall in love with Antoinette, a cousin of a cousin from Charleston, West Virginia. She has a cute Southern accent, which I have never heard before. For the first time I begin to feel entranced, I am literally in a trance with regard to Antoinette. She is the most beautiful being I have ever met. I use every excuse to be with her.

At the end of the summer she goes back to Charleston. During Easter vacation, my family go to White Sulphur Springs, Virginia, and I take a bus to Charleston to see Antoinette. I visit her family, see her again, and am totally disillusioned. All the boys have the same accent she has. Besides, she plays many games, keeping me away from her boyfriend. I finally meet him, a strikingly handsome blond who also has a Southern accent. I leave in shame and embarrassment and give up my first love.

That summer I went to work for the first time. I was fifteen years old and asked Father if I could get a job with the Northwest Airways, of which he was now president.

I went to the manager, Colonel Britton, to apply for a job. He asked me what I could do and I said, "I am an amateur radio operator and would like to work in the radio department." During that summer I met many of the pilots. The head of the radio department put me to work pulling all the wiring on the Hamilton airplanes, putting in shielded wiring to prevent the noise of the engines from getting into the radios. It was a very hot summer, and it was extremely hot work. The airplanes would get to 110 degrees Fahrenheit inside at the airport where I was working. I was paid fifty dollars a month. I spent three months on the job and then went back to school totally disillusioned about what it was like to earn money in that particular way.

I learned that those who knew more, who had learned more in college, were the ones who got the good jobs and the higher pay. I vowed to continue my education and make myself fit for doing the kinds of things that a young scientist should learn to do. I would not be caught in the ignorance which I found among most of the workers in the airlines. I also decided that I would not be the kind of person who was in management in a business.

Father's bank and the idea of working in banks appalled me. It looked like the dullest kind of existence, sitting at a desk and doing repetitious paperwork endlessly. I could not see the power inherent in these operations. I was well acquainted with bank executives and their golf games. I learned to play golf and discovered what shallow conversations took place on golf courses. I preferred studying philosophy and science.

At the St. Paul Academy I took my first science courses and realized that I knew most of the physics and chemistry in those introductory courses. So I spent my time in class devising new experiments, new demonstrations of things which I wanted to see and about which I had only read.

My teacher, Mr. Varney, had a vacuum pump, a plate, and a bell jar and introduced me to the phenomena of gas discharges, electrical discharges in gases at low pressures. We made glow tubes and I began to learn about the various colors of light emitted by gases excited by electricity at low pressure. I learned what makes neon signs emit light. In the pendulum experiments, I set up a photo cell and an electric counter to count the number of times the pendulum swung back and forth each minute. I learned the laws of the pendulum with this experimental setup. I learned the law connecting the length of the pendulum with its vibration period, very rapidly. I stayed after school hours and did further experiments under Mr. Varney's gentle direction and encouragement.

I took the required course in athletics. I played football. I was a guard on the first string. We won the city championship that year. My name appeared in the newspapers as the key person in one of the championship games. I was exhilarated briefly and then realized that today's headlines wrap tomorrow's garbage. I became disillusioned with athletics and decided to do something else.

Mr. Herbert Tibbetts, the master of English and Latin, en-

couraged my interest in philosophy. He referred me to Imman-
uel Kant's *Critique of Pure Reason*. I studied it and studied it. At
first it made no sense, then gradually I realized that with words
and ideas one can prove anything, as long as there is no reference
to experiments in the external world. I saw that the only hope for
science was experimental science, in which one tested one's
ideas by experiment and experience. Kant could prove the thesis
and the antithesis in parallel columns.

My illusion that by pure reason one could arrive at an adequate picture of the universe was shattered. The absolutes that I
had learned in the Catholic Church were destroyed finally and
irrevocably. My search for reality was initiated. Mr. Tibbetts
asked me to write something on reality for the school paper. I
spent weeks on this and struggled with many ideas way beyond
my grasp, but I finally wrote an article contrasting the activities
of the brain and the activities of the mind. I did not know it at
the time, but this article turned out to be a plan for my future
scientific career.

A slight injury to a knee in a football game allowed me to
withdraw from sports gracefully. Mr. Tibbetts then encouraged
three of us to do a documentary movie on school life. We spent
the year shooting with my mother's 16-mm motion picture camera everything that happened on campus, including football
games, hazing in the two school clubs, meals in the cafeteria,
classes, and faculty meetings. We edited our own film after ac-
cumulating six thousand feet of black and white footage. We cut
the film to an hour-long feature.

The headmaster called us in before the film was to be shown
to the parents and the student body. He said, "I want a private
showing of this film before they see it." We objected and said
there were only three days left and that if he had any objections
to the film we wouldn't have time to edit it. He demanded that
we show the film to him, so we did, quaking the while for fear he
wouldn't approve of our work.

We had scenes in which older boys were fighting with
younger ones, some of them getting hurt during these fights. We
had also shown some paddling of pledges to the clubs. He ob-

jected to these scenes, and the oldest of us, Louis Goodman, gave a very articulate, political plea for the film. Mr. Briggs was so impressed with this plea that he did not insist on censoring the film and allowed us to show it the way it was.

As a consequence of the film, the board of trustees had a meeting and abolished the two clubs and their hazing and insisted that there be better relationships between the faculty and the students to prevent the kind of student conflicts which we had shown in the film.

I began to realize the power of recorded events on people and organizations.

At the St. Paul Academy, I became imbued with an ambition to become a research scientist. I took the college boards and was accepted at the Massachusetts Institute of Technology. Mr. Varney had a long talk with me about M.I.T. He said, "There is a scientific school on the West Coast in California, the California Institute of Technology, which has a better science course than M.I.T. I would like you to apply there also. By the way, they do not accept the college boards; they have their own entrance examinations."

I was immediately intrigued and challenged by the fact that they did not accept college boards. Mr. Varney arranged for the examinations to be sent from Cal Tech to the St. Paul Academy. I took them over a three-day period. These were the most difficult exams that I had ever taken and I felt I had flunked them.

Meanwhile, I'd fallen in love with a girl from the East Coast, from Boston. Her name was Amelia.

Amelia was a gentle, cultured, artistic, withdrawn girl. To me she was beautiful, delicate, and at the same time athletically competent. We had many dates under very restricted circumstances. While she was attending school in St. Paul, she was staying with an aunt who chaperoned her very carefully. She had been accepted at Vassar College, so I went east when she left for the summer and visited her on Cape Cod. I met her family and then had my first fight with her over what college I would go to. She refused to promise to date me if I came to M.I.T. I went back to Minnesota, disappointed and heartbroken. When I arrived Mother said, "You have a letter from Cal Tech."

I opened the letter and it said, "We are pleased to inform you that your grades on our entrance examinations were high enough that we wish to offer you a scholarship."

I had an immediate fierce joy. This would settle the case of Amelia, and I would be independent of Father for the first time.

I delighted in showing Father the letter. He immediately disagreed and said that I should go to M.I.T., that it was a far more important school than this unknown Cal Tech out on the West Coast. I argued and fought with him and left in a rage. He had said that he would not pay my fare to California but that he would pay for my tuition, room and board, and transportation to M.I.T. In my disappointment and grief, I got on my horse and rode off through the woods trying to think it through. The next day Mother and Father asked to talk to me at lunch.

Mother said, "Do you really want to go to California to school?"

I said, "Of course. Cal Tech is a much better scientific school than M.I.T. I have never been to California and I want to go there. I don't want to live in Boston or Cambridge. I know what they're like and it's too much of the city life for me. I hear that Cal Tech is in a quiet little town called Pasadena."

Father said, "You know, your mother and I took our honeymoon in Pasadena at the Huntington Hotel."

"Well, all I know is that I want to go there to college. I have a scholarship there and I feel that I'm making my own way independently of you."

Mother then said, "We've talked it over and we'll leave the decision up to you. Your father strongly recommends that you go to M.I.T. I prefer that you go where you want to go."

Thus was the decision made. Finally I would go to the school of my choice, with a good deal of fear of making new friends and of taking on the job which I didn't yet fully understand.

During John's summary of his life from childhood through prep school, Robert remained silent. As John finished there was a long silence.

Robert then said, "There is much you have become aware

of which you had previously denied. Our work continues. Your story is as yet incomplete. Today you were so deeply into the material that I let you go on for two hours. We will continue tomorrow at the usual hour."

John lay on the couch, silent, still immersed in his past life. Finally he got up and left Robert's office, re-engaging in the current external reality.

School Days

courtesy of the Lilly family

Schoolboy John Cunningham Lilly.

courtesy of the Lilly family

First string guard (right)

Cadet (front row, fifth from left)

courtesy of the Lilly family

From Physics to Biology

John entered Robert's office feeling dissociated from himself. He was early for his hour and sat in the outer office contemplating his past life, his decisions, and what had generated him up to the present time.

The analysand preceding him came out of Robert's inner office and left by the street door. In a few minutes Robert came out and signaled John to enter the inner office. Before lying down on the couch John said, "Today I feel separated from my past life. I feel as if I were someone else looking at my past life from some other position than being inside my own head. There is a peculiar dissociation from that growing young man. I feel that today I must speak about John as if he were not me."

Robert: "You mean that you cannot connect yourself with your past, with your past feelings, or with the past events?"

John: "Today it is as if I were not me. That somehow I am a separate being from that human called John. I do not have much feeling about John's past life today. I seem to be riding somewhere above him, watching him."

Robert: "I suggest that you maintain this state of being and speak from it."

John: "I am experiencing some fear at doing this."

Robert: "What is it that you fear?"

John: "I fear that I will never reconnect with John if I am not careful."

Robert: "Who are you?"

John: "I am an extraterrestrial Being supervising my agent John."

Robert: "You are an extraterrestrial Being supervising your agent John?"

John: "Yes."

Robert: "As an extraterrestrial Being, do you have access to John's memories?"

John: "Yes, I do. As John lies here on the couch talking to you, I am above him watching him talk. I control him. I can turn on his memories and help him integrate his past life. Do you want me to do that?"

Robert: "Yes, go ahead. When John last spoke to me he was reviewing his life and finished the review through prep school as he was about to go to college. Do you want that review to continue?"

John: "Yes. I will instruct my agent to give you a continuation of his integration of his past life. Because of my presence and his knowledge of my presence, he will be speaking through me. I, as well as you, will be observing the material that he generates from my point of view, not his own. I will speak through him, recounting the story in the third person rather than in the first person."

Robert: "You mean that you are aware of all of his past life?"

John: "Yes. I have been with him during his whole life and have complete access to all his memories, some of which he does not remember. I believe you use the terms 'repression' and 'suppression' of memories for this process. As John speaks in the third person, I will be inserting portions of the true story insofar as my agent can take it at the present time. As you and I both know, it will be necessary for John to do many such reviews with my help and with your help.

"Are you making any judgments about this way of speaking, of my talking through my agent?"

Robert: "My job is to be objective, not involved, and not to make judgments at this time. Go ahead and speak for John."

John: "I am worried about allowing that Being to speak for me."

Robert: "Let the Being speak through you."

In the following material, John on the couch spoke as if he were a Being from somewhere else, recounting the story of his life from prep school through Cal Tech.

In the fall of 1933 John's parents drove him to the railroad station. He was literally leaving home for the first time. He was eighteen years old and facing a future in which he would only reluctantly return to his home in Minnesota. His home had furnished the nurturance needed for his new life.

The train traveled through Iowa, Nebraska, Wyoming, Utah, and Nevada, and arrived in Pasadena, California, three days later. During those three days he gradually developed a sense of his aloneness, of a future in which his family and his old friends would be less and less important to him inside his own being, even though they would remain important to him in terms of his survival on the planet.

He arrived at the California Institute of Technology as one of one hundred entering freshmen. He registered and was assigned a room in a student house known as Blacker. The first few days were spent in an orientation course given by the YMCA in the San Gabriel Mountains.

At this gathering of freshmen, he learned that his fellow students were like himself, somewhat eccentric with regard to the rest of humanity. When he talked with another freshman, he began to see that he had come home, that each of them knew as much as he about science and were dedicated to learning more within the current scientific knowledge.

He became acquainted with the Cal Tech motto, "The truth shall make you free." His belief in this motto was already formed, as it was among his fellow students. He learned that most of them looked down on the football team, which had managed to lose every game for the last five years. He heard that the athletic program was a requirement, that so-called P.E., for physical education, was considered a joke among the students.

The faculty and the administration expected the students in each of the four student houses and the off-campus students in the Throop Club to establish rivalries. They had their own football games, tennis games, and chess competitions to develop a group feeling in each house.

The freshmen in each house were segregated into their own "alley." The choice of rooms was dictated by seniority. The seniors and the juniors got the best rooms, the sophomores the least desirable, and the freshmen were given arbitrary assignments. The students in each house ate together in their own dining room. There were no undergraduate or graduate women students at that time. Each house had the austerity and the dedication of a monastery.

A member of the faculty lived in each house. In Blacker House the resident was Dr. Harvey Eagleson, a member of the humanities department who taught English literature.

During John's freshman year, Doc Eagleson became a powerful teacher for him. Doc insisted that his students attend teas in his room, and John became acquainted with Freud, with art, with all those things that were not taught at Cal Tech. Doc treated the symptoms of homesickness, loneliness, separation from family with wit and understanding. He considered the students his own family. He had never married and had written a novel about his one love, a woman who had died in an automobile accident. His apartment was lined with a perfect collection of Japanese prints by Hiroshige. Doc attracted those students who were interested in art, philosophy, literature, and writing. He encouraged their literary endeavors. For John he was an inspiration to continue the writing he had begun at the St. Paul Academy.

In his opening course in English literature, Doc started out with the following statement: "In 1859 God died. The pile of dirt upon his grave has been increased continuously since that time."

To John this statement was a shocker. He was still in the throes of his earlier beliefs in regard to the Catholic God. Doc's lecture went on to explain that in the year 1859 there were two important events leading to the demise of the old religious beliefs: the publication of Charles Darwin's *Origin of Species* and the birth of Sigmund Freud.

Thus Doc represented the first person with whom the young John could speak about beliefs, about swinging from Catholicism over to a belief in science and its conclusions about the origins of Man. Under Doc's tutelage John bought the works of Freud and studied them in his spare time. He read Darwin's books and gradually saw the emerging picture of reality as dictated by the current consensus signs. He personally went through the battles that were started in the nineteenth century and revised his own beliefs to bring them up to date in the twentieth century.

John found that he had been assigned to the A Section. There were twenty students in A Section, all of whom had scholarships to Cal Tech. This section had special teachers, faculty who were particularly interested in bright students and who pushed them in their courses. He found that the discipline he had acquired at the St. Paul Academy stood him in good stead, that the study habits, the long hours over books were expected of him in A Section. He found his peers among the other freshmen in his class. He made friends with the few graduate students resident in Blacker House.

He had come to Cal Tech dedicated to becoming a physicist. As he met others who wanted to go into physics, he realized that this was an incredibly demanding discipline. Physics required advanced mathematics and a supreme dedication.

Approximately half the student body were to become scientists, the other half engineers. He found that the engineering students were the ones who played football and were more outgoing, organizing group activities. He shrank from them and devoted his time to the science students.

He had many doubts that he could survive the rigors of training at Cal Tech. He expressed these doubts only to Doc Eagleson.

His roommate was a tense, dedicated mathematics student by the name of Jack Mason from San Diego. As it turned out during the freshman year, Jack was in way over his head. Unable to survive, he left to go to Stanford University. Among the undergraduate body there was a great deal of snobbery in regard to other schools. Cal Tech was at the top of the heap and the students were an elite. This not-so-subtle snobbery among the undergraduates was a mimicry of the more subtle snobbery of the faculty. Cal Tech accepted only the best: the young undergraduate did not perceive that "the best," as defined by Cal Tech, was based on a very narrow set of criteria within the academic disciplines.

In later years John was to realize that the Cal Tech graduates were an effective, operational group of men. Associated with him in Blacker House was a future head of the Atomic Energy Commission. Among the graduate students was a future president of Stanford University. Among his faculty was the future head of the Jet Propulsion Laboratory. One of his teachers was to become the organizer of the yet to be established National Security Council. The faculty also included several Nobel prize winners.

Dr. Robert A. Millikan was the head of Cal Tech. During John's first year he and his wife held dinners for the freshmen. Dr. Millikan's wife, Greta, was a gracious lady in the old tradition. John was to work with their son Glenn during World War II.

During the frantic push of courses in the freshman year John immersed himself in physics, in mathematics, in chemistry for thirty-six hours a week. In addition he took humanities courses: history, English, economics. These were a relief from the pressing business of science. He took the English courses from Professor MacMinn, who allowed him to write essays about what he was reading in Freud's works. MacMinn also encouraged imaginative productions. In 1934 John wrote one essay as an account of World War I from the viewpoint of an extraterrestrial traveling in an orbiting vehicle about the earth. Aldous Huxley's *Brave New World*, published a year or so earlier, was read assiduously by the young undergraduate.

At the end of his freshman year he found that he had passed sufficiently well so that his scholarship was granted again for the sophomore year. As the year ended Doc Eagleson asked John to come to talk to him in his rooms.

Doc said, "I have an important message for you. You have made a basic mistake at Cal Tech and I wish to straighten that

out with you. I do not want your future prejudiced by a continuation of that mistake."

John immediately became fearful and imagined all sorts of
possibilities as to what the mistake was.

Doc continued, "Your mistake was in accepting the scholarship that was given to you. At the faculty meeting about scholarships, the fact that your father is well-off and has sufficient money
to send you to Cal Tech was brought up. You should have accepted
the honor of the scholarship without accepting the money."

John felt a sinking feeling in his belly and thought to himself, They do not understand the real situation. I was attempting
to become free of Father.

John: "Why wasn't I told this? My letter said that I was re-
ceiving a scholarship to pay for my board, room, and tuition,
with no other explanations. It didn't mention any conditions for
accepting the honorarium. I feel the faculty is being unjust."

Doc: "Unjust or not, that is how the faculty thinks. That is
the way the human reality operates."

John: "The Cal Tech motto is, 'The truth shall make you
free.' You mean to say the truth is not communicated in full here
at Cal Tech?"

Doc: "Truth is relative. Scientific truth is one thing; human
truth is another thing. The human truth here is that your reputation with the faculty has been damaged by your accepting the
scholarship when your father could pay."

John: "Doc, you know about my family, that I accepted the
money to be independent of my father. I am incredibly disappointed that Cal Tech does not back me up in this and that they
tell me in such a sneaky way."

Doc: "Calm down, John. This isn't the end of the world. If
you want to, you can accept the money for next year. I am
merely telling you that if you do the faculty will be prejudiced.
Their judgments about you in the future may spoil some of your
chances to do other things."

John: "What the hell is really going on, Doc?"

Doc: "It has come to the attention of the faculty that your
father is wealthy enough to become an Associate of Cal Tech.

Do you know about the Associates program?"

John: "No."

Doc: "Tuition does not support the undergraduate program. Cal Tech needs additional money, so Dr. Millikan has worked out the Associates program. Associates are wealthy men who contribute ten thousand dollars or more. They are told the aims of Cal Tech and have special days when they tour the campus. Those in charge of the program have found that your father is in a position to contribute money to Cal Tech. They cannot approach him if you are a scholarship student, whether your funds are donated by the Associates program or not."

John: "Look, Doc, one of my major triumphs in coming to Cal Tech was that I could be free of my father's influence. If he becomes an Associate here, my fellow students may think that somehow or other my father is buying my way through Cal Tech."

Doc: "Well, it's high time you learned the realities. There's not much more I can say about this, except the faculty counts on many sources of income to support their teaching and research— donations to Cal Tech, consultancy fees with industry, grants from corporations or government agencies who want certain kinds of research done. Your picture of how science is paid for is incomplete. After you graduate, after you've completed your education and you're either on the faculty of a university or working in some laboratory, you may have to ask your father to contribute to your own research.

"I suggest that you make peace with your father and gain his goodwill. The illusion that you are operating independently of him is a luxury you can no longer afford. You should develop some of the talents of Machiavelli. Read his book *The Prince*. The human reality is not a scientific reality. Power and money representing power are the way it operates. Times have not changed that much since Machiavelli."

John: "You mean, then, that Cal Tech operates the way every other institution of higher learning operates, that it is supported by the seduction of power, by charity, that it does not earn its own way?"

Doc: "John, you can't be so emotional. Take a course in economics next year and learn something of what your father really does rather than harboring some romantic illusion based on your parent/child conflicts. The flow of money and power supports physical research and biological research and all those things which interest you. In addition to your scientific and technical know-how, you must learn where to go to get support and how to talk to the people who can give it to you. You carry around a hostile model of your father. I suggest that you analyze your conflict with him and stop projecting that model onto Cal Tech, onto me, and onto your future supporters. Someday you may need psychoanalysis to straighten this out.

"Meanwhile, I think you should tell the committee that you will accept the honor of the scholarship but not the money. Think it over this summer and let them know your decision next fall."

John: "Well, right now I feel as though I don't even want to return to Cal Tech. I would rather drop out and go to work and become really independent of my father."

Doc: "John, that would be very foolish. You are in the midst of one of the best science educations in the world today. I'd advise you to come back next year and continue the training which you very badly need. You are only nineteen years old and have much to learn in the scientific reality and in the support of scientific research.

"You have an admirable independence of mind; however, this can be a liability as well as an asset. You must realize that independence of mind has the reality of interdependence with other humans. You must study this interdependence even as you study your physics, your biology, your chemistry. Eventually you will realize that there is no such thing as an independently operating human; there is only the illusion of independence."

That summer John returned to Minnesota, rode horseback, lived with his family, thought about what Doc had said, and finally had a conference with his father in which he explained what Doc had said to him and asked his father for support. His father and mother talked it over and informed him that they would support him during his education, that they would set aside the money for this in a trust fund.

John wrote a letter to the Scholarship Committee asking that the money for his scholarship be put back in the fund and distributed to more needy students.

Toward the end of the summer, his father informed him that he had been invited to become an Associate of Cal Tech and that he was contributing a thousand dollars a year for ten years to the Associate funds.

That summer in Minnesota John was made even more aware that he was an eccentric. His friends from the St. Paul Academy had gone to the East Coast to Ivy League colleges. When he met them that summer he realized that they were further and further away from his particular universe of experience. Their stories of college life were so alien to his experience at Cal Tech that he was unable to understand how anyone could function in the ways in which they functioned. Most of their stories were about football games, about parties, about whom they met, whom they had talked to, and the gossip of the East Coast. His high school friends had gone to Yale, Harvard, Princeton, Vassar, Holyoke, Bryn Mawr, Smith. Their dates, parties, trips to New York to see the latest show on Broadway or to the Yale-Harvard game were the content of their conversations.

John felt out of the picture, separate and quite lonely. The parties he attended were boring, the girls beautiful and unavailable. He felt like an outsider looking at a scene in which he could not participate. Amelia had returned to Minnesota that summer. She was remote and did not want to date him. He dated another girl from his high school days and found her shallow and unsympathetic to his discussions of philosophy, science, and Cal Tech. He found one other Cal Tech undergraduate in Stillwater, Minnesota, and spent his time with him.

He began to look upon his hometown of St. Paul as a dead end, as a place where there was no scientific curiosity, no curiosity about the world, about the universe. His conflicts with his family sharpened, and he decided that he would never again come back to Minnesota if he could avoid it. His remaining bonds with his family, with his traditional life, were gradually dissolving. He increasingly resented the necessary financial dependence on his father and mother. He spent most of his time in the radio station talking to radio hams around the world. Everywhere he went in Minnesota, he was faced with questions about his father, rather than about the work he wished to discuss.

In the fall of 1934 he returned to Pasadena, to Cal Tech, and took up his life as a student once again.

During that year he was exposed for the first time to biology. As a major in science rather than engineering, he was assigned the elementary course in biology under Thomas Hunt Morgan. Professor Morgan taught all science students this course. It shaped John's future as no previous course was able to do.

In the opening lecture Professor Morgan showed a slide of an embryo in the uterus.

He said, "This is the embryo of a pig. No, this is an embryo of a monkey. Oh, I'm sorry, I've mixed up all the slides. This is an embryo of a human. Oh, well, at this stage it doesn't make any difference; they all look the same."

To John this was a shock. That all mammals went through similar stages of development, including Man, was an entirely new concept to him. The embryological evidence for the origins of Man as one among many mammals was an exciting idea to the young student.

As the course progressed he decided that he had found the area of science in which he wished to learn more. During the subsequent years at Cal Tech, he shifted to biology and found the most interesting area for him was the study of the origins, the development, the functions, and the structure of the central nervous system, the brain.

There were three students specializing in biology as undergraduates. Each of these students was taught by ten times as many faculty as there were students. Each undergraduate did a research problem in one of the biological sciences. Each student was expected to do genetics research on the fruit fly, which had given Professor Morgan the opportunity to map the genes along the chromosomes of the fruit fly. He received the Nobel prize for this work. John took courses in genetics, in plant physiology, in vertebrate zoology, in biochemistry, in neurophysiology, in embryology, in mammalian anatomy.

He became intrigued with single-celled animals, the protozoans. Under the microscope he watched algae move around rapidly, absorbing light through their chlorophyll and whipping their busy way through the water with their single flagellum or their cilia. He watched the development of plants from seeds; he watched the development of sea organisms from their fertilized eggs. He dissected cats, frogs, fish. He studied the brains of the smaller organisms and the brain of the cat. He saw demonstrations of the electrical activity of the brains and nervous systems of crayfish, of cats. He studied the literature of neurophysiology and became intrigued with the electrical activity of the brain and the possibilities inherent in the various techniques used.

In his first undergraduate faculty seminar John discussed a paper by Edgar Adrian, "The Spread of Electrical Activity in the Cerebral Cortex." This paper excited his interest to the point where John decided to devote himself totally to the study of the brain and its electrical activity. He conceived of a method of portraying, of visualizing, the electrical activity of the cerebral cortex in a television-like manner to see the waves spreading rapidly over the surface of the brain. Many years later he was able to carry out this idea, build the apparatus, and see these waves as they had never before been seen.

During his junior and senior years, he was asked to join the Anaximandrian Society in the biology department at Cal Tech.

This society was sponsored by Dr. Henry Borsook, the professor of biochemistry. In the society each of the students was asked to research a paper on certain subjects within biology and medicine. John chose the history of man's ideas about the brain. He researched the literature from the time of Hippocrates, Aristotle, and Galen, up through the nineteenth century. Researching this paper expanded his interest in the physiology of the brain and the development of ideas about the connection between the brain and the mind.

In various conversations with Professor Borsook, it finally became apparent to him that he would dedicate the rest of his life to research on the mind and brain. Dr. Borsook said, "The kind of knowledge that you need to pursue research on the mind and the brain is available only in medical schools. You will find that unless you take an M.D. degree you will not be able to do the kinds of things you wish to do in your scientific endeavors. I suggest very strongly that you take the M.D. degree rather than a Ph.D."

John took this advice and applied to several medical schools.

As the extraterrestrial Being finished speaking, John took over his body once again and lay on the couch silently contemplating what had happened during this hour.

Robert: "To whom am I speaking now?"

John: "The Being has left. Somehow I am me again."

Robert: "While the Being was speaking, were you aware of what was going on?"

John: "Yes, but every so often I was panicking, afraid that I would not be able to take control of this body again."

Robert: "Have you ever had an experience like that before?"

John: "I seem to remember experiencing this kind of 'dissociation' when I was very sick as a child. There were long periods when, lying in bed, unable to go out and play with my brothers, I felt as if the Being were present, speaking to me and at the same time through me.

"There were other times, but they were much briefer. Coming out of anesthesia from a tonsillectomy, being put under nitrous oxide to have four wisdom teeth pulled, falling out of a tree on my head in grade school."

Robert: "So you've experienced this Being before under conditions in which your physical organism was depleted by disease, fever, or anesthetics?"

John: "Yes, but today wasn't like that. There has been no disease, no injury, no anesthesia, no drugs."

Robert: "So you are becoming aware that you tend to split into human being and extraterrestrial Being?"

John: "Yes. I now immediately think of all sorts of psychiatric diagnoses of such states. I think of hypnotic regression and splitting. I think of psychotic episodes in which a person develops two personalities.

"Apparently I am generating all these explanations for you as a psychoanalyst. Somehow these explanations seem unreal to me."

Robert: "So your internal reality rejects psychiatric and psychoanalytic explanations for what has been happening to you here today."

John: "Yes. There seems to be something inadequate in these ready clinical explanations. But on the other hand, my running away from those explanations may be merely another attempt of my very ingenious unconscious to explain away by other means, by more spectacular means, what is really going on inside me."

Robert: "How would you explain what has been going on here today?"

John: "Well, all I can say is that this seems beyond anything that is explainable by my present knowledge of science, psychoanalysis, and psychiatry. Somehow, something is going on that surpasses our present framework of thinking and explaining."

Robert: "I am not here to argue explanatory theorems. I am here to point out to you what is going on inside you insofar as I am aware, and can become aware, of those processes, of you as you really are, not as some clinical diagnostic tools would label you. Your private explanations are grist for the mill of psychoanalysis. My own beliefs about what is going on are irrelevant.

"Let us continue tomorrow at the usual hour. I suggest that this week we devote seven days rather than five days to your analysis. I suggest that you come back Saturday and Sunday in addition to the rest of the hours this week."

John: "That suggestion brings up a feeling of fear. It seems to me that you were frightened by what happened here today and now want to be able to control me and supervise what happens in more detail during a difficult period in the analysis."

Robert: "Whatever. Let us continue tomorrow at the usual hour."

First Marriage
and Medical School

7

The following day John came into Robert's office and lay down on the couch with an energetic thump.

John: "I have been thinking about what happened here yesterday during the hour, with the Being taking over and recounting my history as if from a third-person viewpoint. The reality of that experience is still with me. If all this is happening just inside my head and there is no real, mysterious Being from some extraterrestrial place, then I was using this as a means to escape involvement in the feelings of my past life. In that case I was simulating the Being to avoid penetrating more deeply into my own unconscious. I was pretending to be a historian, giving the history of somebody else, not me. I find with this method of speaking in the third person that I can penetrate some things I can't penetrate when I speak in the first person. I find the Being a convenient construct which allows me to sit outside and observe what is said as if I did not own it myself.

"Today I want to continue in that mode of discourse. This will allow me to check out various events of my past life from a more objective standpoint."

Robert: "So you feel safer speaking as if your past life belonged to somebody else?"

John: "Yes. Today I will continue the review of my life as if

I am a historian, writing it in a book. Rather than an autobiographer, I will be a historian."

Robert: "Is the Being real to you today?"

John: "Not as real as yesterday. He, she, or it is somehow not in this room but is more remote, somewhere else, as if listening to me speaking here to you."

Robert: "It makes little difference to me how you speak, what grammar you use, whether the first or the third person."

John: "OK. I still want to speak in the third person."

For the first two years of his education at Cal Tech, John had lived a monkish existence, dating infrequently and working very hard at his education.

In the beginning of his junior year, at the age of twenty, he met Mary. He dated her several times and in his naïve, monkish way fell in love with her. His Catholic upbringing was still determining how he acted with respect to women. He still believed that he should not have sexual intercourse before marriage. He decided that when he was twenty-one and legally free of parental direction he would marry Mary.

His work at Cal Tech began to deteriorate. He did not get enough sleep and he did not apply himself to his work the way he had the previous two years. He became exhausted. He contemplated leaving Cal Tech and going to some easier college or university. In February of his junior year he went to see his parents and told them that he couldn't continue. They sent him to a neurologist who advised him to drop out of school and find a physical job, forgetting the intellectual work for a period of some months. With the help of his father, he obtained a job in a lumber camp in Oregon. During the winter he worked in the lumber mill on the dry chain, sorting lumber into various compartments and moving it from the dry chain to the compartments assigned to him. When the snow left the woods, he joined the survey crew that was laying out railroads to haul the trees out of the woods.

In January before he left Pasadena, on his twenty-first birthday, he asked Mary to marry him. She accepted. They became engaged.

While on the survey crew, John lived for a time in the woods with various other crews. He was assigned to a bunkhouse in which there was an epileptic. Once he was awakened in the middle of the night by animal-like sounds that filled the room. Another member of the survey crew got up and went over to the epileptic, who was having an attack, and inserted a spoon in his mouth to prevent him from biting his tongue. For a while John was frightened by this episode and moved into the nearby town. He bought a copy of *Gray's Anatomy* and began to study it in preparation for medical school.

John was the chief brush cutter for the survey crew. In pursuing a survey line through a swamp, he was cutting brush with an ax and badly cut his foot. He thought at first that he had cut the head off the small dog which attended the survey crew because there was no pain. When he realized what had happened, he stopped the blood gushing from his foot by lying down in the swamp, raising his foot in the air, and holding on to the artery. He shouted for help and the survey crew took him to the hospital.

The surgeon was sewing up his foot. He said, "You must know something of anatomy. That ax went into your foot in such a way as to hit nothing important and yet go in very deeply."

John, in a light manner, said, "I've been studying *Gray's Anatomy.*"

Surgeon: "Are you planning to go to medical school?"

"Yes, when I graduate from Cal Tech, I am going on to medical school."

Surgeon: "You'll have to stay here in the hospital until this heals. I will be giving you typhoid vaccine to raise your temperature to take care of possible infection in your foot. Here you will learn a lot that will stand you in good stead in medical school."

Under the influence of the typhoid vaccine his fever went up to 103 for a period of three days. His foot slowly healed.

He was in a ward with automobile-accident cases, with mill-accident cases. He could not sleep because of the noise, the moans, and the cries of the other patients at night. His resolve to go to medical school and to be of direct help to such people increased.

When he recovered he left Oregon, went back to Minnesota, and made plans for his marriage.

His father and mother insisted on a Catholic church wedding and obtained the services of the priest who had been influential in John's earlier education, Francis Thornton. The Catholic church chosen was in Pasadena. The wedding party flew out from Minnesota to join Mary's family and friends in Pasadena. It was a formal wedding with the men dressed in morning coats. The solemn ritual of the Catholic Church dominated the scene.

Mary, a non-Catholic, was given Catholic instruction and agreed to bring their children up in the Catholic Church. After the honeymoon John and Mary found a small apartment near Cal Tech and began their married life as students.

John was the only married undergraduate at Cal Tech and felt separated from the student body by his marriage. Mary returned to art school and John continued as a junior at Cal Tech. He was able to resume his intense work.

During that year Mary found she was pregnant. Late in her pregnancy she fell and broke her spine. She was in very bad pain during the last months of the pregnancy. It was decided that she would return to Minnesota to be taken care of by John's parents. The baby was born while John was taking final examinations for his junior year. Meanwhile, Mary's pain increased in her lower back and legs.

During the summer of his junior year, John returned to Minnesota and, on his father's recommendation, went to the Mayo Clinic and talked to Dr. Will Mayo about what medical school he should attend.

John waited for Dr. Mayo in his outer office. He had been up late the night before and fell asleep. Dr. Will came out of his inner office, tapped the young man on the shoulder, and said, "Do you want to see me or do you want to continue sleeping in my outer office?"

The embarrassed young man got up and apologized. In the subsequent conversation Dr. Will said, "There is only one place for you to go to medical school. The basic course that you will have to master is human anatomy. No matter what you do in the future in medicine or in medical research, that is the course that you must absorb. There is only one place in the United States where that course is taught adequately. That is at Dartmouth Medical School in Hanover, New Hampshire. Dr. Frederick Lord teaches a course that is not duplicated anywhere else. Go to Dartmouth and start your medical school."

By coincidence John's younger brother, David, was an undergraduate at Dartmouth. Dr. Will's advice and the presence of his brother made him decide to go to Dartmouth. He sent in his application and was finally accepted under rather unique conditions.

Dartmouth rarely accepted medical students from outside of Dartmouth. The undergraduate students took their first year of medical school as their senior year in the undergraduate college. John sent an application to be admitted to the medical school. Dr. Will sent a letter to the dean, Dr. Bowler, recommending the young student.

John had learned Doc Eagleson's lesson well. This time he was not loath to use influence to move on the path he had chosen.

In John's senior year at Cal Tech, his father drove his car off a bridge and dropped a hundred feet to the ground below. John went to Minnesota and sat by his father's bed for three weeks while he was in coma in an oxygen tent.

One morning his father came out of coma, looked at John, and immediately said, "You are not going to Dartmouth Medical School, you are going to Harvard."

"I am going to Dartmouth. Welcome back. I am glad that you are out of coma."

John spent another week and returned to Cal Tech to finish his senior year, take the final examinations, and graduate.

His wife and child were at his graduation.

Mary and John decided to drive to Dartmouth with all their possessions. During John's first year of medical school, Mary's pain was such that it was decided she must have an operation on her lower back. She went through a spinal fusion and was in a cast from her pelvis to her shoulders for the next three months.

At Dartmouth John began a whole new series of studies which seemed very far from the pure science of Cal Tech. He found, as Dr. Will had predicted, that Dr. Lord's course in anatomy was extremely thorough. He spent several hundred hours dissecting human cadavers. He thought bacteriology fascinating. The histological sections of human tissues to him were evolutionary works of art.

John took up skiing in earnest and worked out a method of determining the position of the center of gravity of the skier in various body movements. He worked out a method of recording the sounds of the human heart in parallel with the electrocardiogram.

Among the twenty-two medical students in his class, there was only one other who did research, Fred Worden. In their two years at Dartmouth together, they laid the foundation for a lifelong friendship.

During these two years Dr. Rolf Syvertsen became the dean of the medical school and advised the young medical student to go to the University of Pennsylvania Medical School rather than the Harvard Medical School, which the young man preferred. Years later he appreciated this advice and realized that the decision was a good one, that he was exposed to certain influences at the University of Pennsylvania that he would not have experienced at Harvard. Somehow, coincidence kept him away from Harvard, first as an undergraduate and then as a medical student.

At Dartmouth he firmed up his resolve to continue his medical education and to go into medical research on the brain and the mind. His courses in neuroanatomy and neurology convinced him of what he had to learn in order to work on the brain versus mind dichotomy. In his introduction to psychiatry at Dartmouth, he discovered the limitations of this field of medicine. His earlier work in neurophysiology at Cal Tech guided him in the choice of directions he could take. The Dartmouth experience reinforced his choice and began to give him the background he would need in the future for his scientific research.

After their first year at Dartmouth, Mary decided to go to Hawaii and take their young son with her. While she was away,

John returned to Cal Tech and did his first scientific research under the direction of Dr. Borsook. In this research he analyzed his own urine and blood and followed the excretion of single products of metabolism in a protein-free diet. He also took test doses of amino acids and followed the excretion of the product. In Hawaii, Mary took a course in painting, learning about the use of pigments that would last.

During his second year at medical school, in the midst of his clinical studies at Dartmouth, John did research on the side. He worked out a method of determining the melting points of drugs in very small amounts on a small heated wire under a microscope.

While John was doing research evenings, he met a surgeon who was attempting to do research and carry on clinical work simultaneously. He advised John not to split himself between the two but to decide which way he was going to go: into therapeutic medicine or into medical research. At this point John decided to take the way of medical research, continuing as a medical scientist rather than a therapist.

John met Professor H. C. Bazett as a junior medical student at the University of Pennsylvania. He told Professor Bazett that he wanted to do research and was given a room in the medical school to carry out this work. Bazett wanted a means of recording blood pressure which was continuous and which could be used on the tilt-table and in flying aircraft. He was English and England was involved in World War II. Bazett was already doing military high-altitude research.

John looked at the machine that had been devised, a photoelectric method of recording blood pressure, and decided that it would not hold up under the stressful conditions which were required for the work. He thought of a new method and proposed it to Bazett, and Bazett furnished the funds to develop the method. During the next year John worked out the method and tried it on himself and on Bazett in the main artery of the arm. The method worked and Bazett took the machine with him to Toronto to use with the Canadian Royal Air Force research unit. Within a year John was encouraged to publish his first solo scientific paper entitled *The Electrical Capacitance Diaphragm Manometer*, describing this method of recording blood pressure.

During this research John met Britton Chance, who was in charge of the department of biophysics. Chance advised him about various difficulties that he was having with electronic circuitry and helped him to solve these problems.

Upon graduation from medical school in 1942, John was asked to join the department of biophysics by Detlev Bronk, who had heard of his work through Britton Chance. For the next eleven years, he remained on the faculty of the University of Pennsylvania Medical School, working under Bronk and then under Chance.

During all this time his marriage continued on its rocky road. The young man was spending most of his time on his student work and then in scientific research. He struggled to partition his life more equably but did not succeed. He was driven to do the research and hence did not spend much time with his family. After the war he had his first affair with another woman.

His parents heard of this affair through Mary and called him back to Minnesota. His father, extremely angry, told him that he would disinherit him if he did not live up to his marriage vows.

Upon his return to Philadelphia, John went through an extended period of fear and paranoia, separated from his family and from his scientific colleagues. He appealed to a fellow graduate from the University of Pennsylvania Medical School, who suggested very strongly that he go into psychoanalysis. John started a search for possible analysts and finally found one, Dr. Robert Waelder.

Robert: "What are your feelings now about the technique that you are using of recounting your history as if you are a historian?"

John: "It seems to me that it is a very useful technique to speak in the third person. Material comes up which wouldn't if I spoke in the first person. And yet I have doubts about it. I do not seem to be so totally immersed in the memories as if I were speaking in the first person. There is not the engagement of my feelings with what I am saying that there was earlier in the analysis when I was speaking in the first person."

Robert: "Our time is up for today. I will see you tomorrow."

John Lilly on the air, speaking into radio microphone.

Young Man Lilly

*Father, Richard Lilly, with his two surviving sons John (left)
and David (right).*

courtesy of the Lilly family

courtesy of the Lilly family

John Lilly

Mary (Crouch) Lilly

John and Mary on an outing.

courtesy of the Lilly family

Electronics Connecting
the Human Brain
and the Human Mind

Robert: "For the last two sessions, you have been an extraterrestrial Being and a historian of yourself, speaking in the third person. Where are you today?"

John: "Today I am a bit insulted by your talking of the Being and the historian."

Robert: "What do you feel is the origin of this feeling of insult?"

John: "Today I am inside myself. There is no extraterrestrial Being and no historian. It's almost as if the last two days were some sort of a play written by somebody else."

Robert: "And your feelings of rage?"

John: "I am disappointed in myself. My 'Calvinistic conscience,' as you call it, says that I am crazy to be talking like an extraterrestrial Being or like a historian. The third person is somehow or other unreal."

Robert: "So your feeling of insult, your feeling that I insulted you with my opening remarks, is a projection?"

John: "Yes, it is merely me, split in regard to my feelings and my conscience. Anyway, all of this seems very trivial and I would like to get on to something more important. I have a secret mission that I wish to explain to you today.

"Oh, that's the reason that I was projecting negative feelings onto you. I haven't shared this secret mission with anyone else, and I had my doubts about sharing it with you until that negative feeling came out and I could look at it."

Robert: "What is your secret mission?"

John: "In order for you to understand it, I will have to give you some background, and I would like to do that as the historian giving it to you in the third person. This will be so that you can understand the basis for that mission as objectively as I can give it to you. So temporarily I will return to the role of historian."

Robert: "Whatever."

At Dartmouth and later at the University of Pennsylvania Medical School, the young scientist continued to learn the dogma taught about the human brain and mind in American medical schools. His courses in psychiatry, in neuroanatomy, in neurosurgery, and in neurology showed him the consensus medical reality, the simulation of the medical men themselves, of the human race in general. He took these years as preparation in what is known and what is simulated about himself and his fellow human beings.

In medical school he saw the fragile, pink, pulsating surfaces of human brains at operations. He listened to and elicited the human sufferings in the patients whom he treated as a medical student. He saw patients suffering from neurological diseases, brain injuries, psychological trauma. He became swept up in the high-altitude research during World War II. After the war, he took a course at the University of Pennsylvania in nuclear physics on How to Build an Atom Bomb.

During these years he lost contact with the Beings directing his mission. He called such thoughts science fiction. The human consensus reality took over his mind, his body, his brain, his social relations.

At the University of Pennsylvania from 1940 to 1945 he did military high-altitude research under the Committee on Medical Research of the Office of Scientific Research and Development on contracts to the University of Pennsylvania Medical School. In addition to devising new techniques for quantitatively record-

ing human respiration at high altitudes and continuously recording blood pressure under difficult conditions, he worked out a
new, extremely rapid (0.01 second) method of determining the
amount of nitrogen in the air inspired and expired by pilots at
high altitudes. Because of this work he was placed on the preferred list of scientists of the War Manpower Commission.

During these years he learned many things about his own
functioning in extreme states. He put himself through explosive
decompression of pressurized aircraft cabins. He experienced
anoxia, the lack of oxygen in the air at high altitudes. He experienced decompression sickness at equivalent altitudes of 38,000
feet in altitude chambers for several hundreds of hours.

He learned about the wartime support of university research by governmental agencies. He was supported by government contracts with the University of Pennsylvania through the
United States Air Force and the Office of Scientific Research and
Development. He learned about war, about the organization of
people in war; he saw that scientific research was forced to become pragmatic, task-oriented, during a war. He saw his physicist friends disappear into the Manhattan Project.

He began to understand that Man's knowledge of his own
mind and brain was at a primitive stage of development and had
very little influence over the power politics, the economic structure of the country, the media, or the new United Nations. He
began to see that his mission might be premature in the human
society on this planet. His desire to study the brain and the mind
of Man was put in abeyance by the war.

After graduation from medical school in 1942, he did his
war work in the Johnson Foundation under Detlev Bronk. The
day the war ended Professor Bronk came into John's laboratory.

Det: "John, the war work is finished. It is time to start thinking
in terms of new research, not war-oriented." Det was looking out of
the window across a low building belonging to the medical school.

He said, "See that building there? I would like to build another story on top, connected to this building. It will cost approximately a hundred and fifty thousand dollars. I suggest that
we approach your father to donate money for that building. If

you can do this, then you will have a secure laboratory space for your own work under my direction."

John: "What sort of research are you thinking of?"

Det: "I would like you to be on my team investigating single neurons. I would like to continue the work that I stopped during the war."

John: "Well, Det, now that the war is over, I would like to resume the kind of research I have visualized since 1937. I want to work on portraying the activity of the intact, unanesthetized brain, over its surface and in its depth, using a television-like method."

Det: "If you wish to do your own research, not on my team, you will have to raise the money yourself."

John: "I don't know that I can raise the money, but if I can, may I maintain a position in the Johnson Foundation?"

Det, obviously disappointed: "If you do your own research independently of me, I can give you a position and a small salary, that is all."

John, feeling very anxious and quite upset: "OK, Det, I will try to raise the money for my own project."

John flew to Minnesota and talked to his father, who agreed to give him $10,000 for his own research from his foundation. He would transfer immediately $10,000 to the University of Pennsylvania for the purposes of brain research to be directed by John.

Using this money, and with the help of the personnel of the Johnson Foundation, John developed a twenty-five-channel television-like display device for the brain's activity on the surface of the cerebral cortex. When the device was completed, waves of activity were seen moving across the cortex in rabbits, cats, and monkeys.

John learned neurosurgery and how to build electrode arrays and the necessary electronic equipment. He saw that the more skull was removed from an experimental animal, the longer it took the animal to recover from the operation. He recognized the essentially fragile nature of mammalian brains in the living state. He found that anything one did to the skull and brain caused injury to the brain, sometimes only microscopic injury, at other times massive injury.

He learned how to stimulate monkeys' brains in the un-anesthetized state. He learned that electrical currents improperly regulated eventually damaged brain. He worked out a new electrical wave form that did not damage the brain of itself. He mapped the movements of the body of the monkey through stimulation of the sensorimotor cortex.

Meanwhile, he learned more about his own mind during his psychoanalysis. He discovered the social realities of being in scientific research. He learned the dictum, "Publish or perish."

His studies of his own mind and of the brains of animals and humans in the forties and the early fifties emphasized the current dichotomy of points of view about these two entities, the brain and the mind. He conceived of a new technique to learn the intimate connections between the brain and its contained mind. In his work in neurophysiology, he proposed the new technique in an obscure paper.

The scheme was somewhat as follows: If one could devise extremely small electrodes and wires to be inserted into a brain in numbers approaching ten thousand or more, one could pick up the electrical activity throughout the brain, record it, and at a later time feed it back into that brain through the same electrodes. One would record the behavior of the animal during the initial record and then during the playback into the brain from the storage (on film or on magnetic tape) and compare the behavior in the two cases. He hypothesized that if the behavior in the two cases was identical, the mind was contained in the brain. If the behavior was not identical, then there was at least the possibility that the mind was not contained in the brain.

He visualized performing this experiment on himself so that he would have the additional source of information of his own inner experience.

Robert: "So you want to hook your brain up to a recording system and then play that recording back into your own brain. Is this correct?"

John: "Yes. I visualize ten thousand, a hundred thousand, a million electrodes inserted through my skull into my own brain, hooked up to an adequate recording system. The records would be made while I am actively doing something. At a later time they would be played back through the same electrodes. I would then be able to tell if I went through the same actions, also recorded on motion picture film, that I had gone through originally. I would also be able to see whether or not I had the same inner experience that I had during the original recording."

Robert: "Is this technique feasible at the present time?"

John: "No, all the methods have to be worked out, including the insertion of the very fine electrodes and leads into the brain, with a minimum of injury."

Robert: "I can see that this raises a number of intriguing questions scientifically. It would enable you to do experiments for the first time on the correlation between the consciousness of a person, his motivations, and the origins of these motivations within the structure of the brain. I appreciate the scientific and philosophical questions which such experiments could answer; however, here we must analyze your motivations for doing such experiments on yourself. What are the dangers to you of such experiments?"

John: "If I can make the leads and the electrodes small enough, and if I can make the necessary holes in the skull small enough, I can minimize the damage and maximize the safety of the procedures."

Robert: "You admit, then, that the present techniques are far too damaging to be used on yourself. Correct?"

John: "Yes, I will have to devote several years of research to methods which are less damaging than the ones that have been devised to date. The plans call for much smaller leads, much smaller electrodes, much smaller holes in the skull than have currently been used.

"I don't believe you appreciate the full scientific value of such work. A human subject so equipped could answer major philosophical questions which have bothered thinkers for centuries. For the first time we would have a close correlation among the four areas: the electrical activity of the brain, the control of the electrical activity of the brain, the concurrent subjective states of the inside observer, and the external behavior, including the vocal output. We would have an answer to the questions, Can electrical currents totally control the brain's activities? Can electrical currents, put into the brain in suitable patterns, totally control the mind in that brain, the observer in that brain?"

Robert: "I appreciate the scientific mission you have set yourself. You have said previously that you had the idea of doing this quite early in your career. You reported that at age sixteen you wanted methods for correlating the activities of the brain and the mind. You saw the possibilities of doing this in 1937 at age twenty-two. Now you consider working on this program full time. I appreciate your scientific dedication and I appreciate the importance of the questions which you are raising. However, I am a bit skeptical about your wish to do it on your own brain. It seems to me that you have not fully evaluated your own motivations."

John: "My motivations are not self-damaging, suicidal, or masochistic, at least they are not consciously so."

Robert: "Well, as you have learned in this analysis, you do have strong components underlying your conscious motivation which could lead in the direction of self-damage. It seems to me that you've picked the ultimate in self-damage by means of tens of thousands of brain electrodes."

John: "It could be. When I look upon myself as a human functioning in the human reality, which ignores totally brain structure, brain function, and the functions of the mind, the criteria of self-damage are obvious. Yes, unconsciously maybe I do want to eliminate myself from the human competitive sphere. But I also detect a tendency to want to be a hero to other humans. A hero takes risks. A hero may end up damaging himself. A hero can fail and become, in his own mind at least, a martyr."

Robert: "Your Catholic background postulates that you worship a martyr, Jesus Christ. As we know from your history during your psychoanalysis, you had many wishes in your early life to become a priest, to become Christ, to hold yourself up as a human willing to sacrifice himself in the service of other humans. Do you consider these realistic motivations for a mission?"

John: "Being a martyr publicly is seductive. It is also dangerous and unrealistic. Martyrs somehow have gone out of fashion. Missionaries have gone out of fashion in an obvious public way. Our methods today are more subtle. One controls the educational system of young children and prejudices them into a human consensus reality that is desired for the control of large populations."

Robert: "As you know, I spend a good deal of my time on public affairs. My avocation is political science. I say these things which you already know in order to emphasize the essentially political and social aspects of your present mission. This is an area which you have neglected in your education and in your research. You have shown tendencies in your relations with your father, in relations with your scientific colleagues, to ignore political realities. What do you think the impact on your scientific colleagues would be if you revealed this wish to insert electrodes in your own brain and learn what the relation is between the mind and the brain by these rather damaging techniques?"

John: "Suddenly I feel that if I did this it would be highly seductive for other persons. It would be intriguing to them. It would set me up as 'the first man to find out about the relations between the brain and the mind.' I visualize myself as a hero, obtaining the Nobel prize. I would go down in history no matter how the experiment turned out."

Robert: "You can see some of the childish aspects of what you intend to do. You wish to gain power, to become well-known. It seems to stem from a need for acceptance by your fellowman. It seems to arise from a desire to exhibit your own mind and your own brain publicly. The secret nature of what you've been pursuing, of what you call your mission, seems to be some sort of a science-fiction script which you keep to yourself."

John: "I am afraid that if I analyze all this I will stop my scientific research and go to something else. I don't want to analyze this anymore today."

Robert: "What is your fear?"

John: "My fear is that my mission is motivated by unrealistic, imaginary needs dictated by my childhood, my loneliness, my inability to love, my essentially inhuman or nonhuman unconscious."

Robert: "As you know, I do not worry about what you have

made conscious. I only worry about what is yet in your unconscious. 'Where id was, there shall ego be.' You know about the ego and its defenses as presented by myself and by Anna Freud. The boundary between your consciousness and the unconscious motivations is all that we can deal with here in your analysis. In the future it will be necessary for you to work out how important this project is to you and to science in the human consensus reality. No one is ever completely free of unconscious motivations; however, it is possible to make more and more of those motivations conscious and to deal with them in the light of day."

John: "Well, as usual, you've kept me from being too impulsive about something based on unconscious motivations. I intend to pursue this program and, if possible, to devise minimally injurious systems of exteriorizing the electrical activity of the brain and of restimulating the brain through the same means. Then eventually I will use myself as the subject of the experiment. I have been brought up in a school of human physiology in which one's self is the obvious first subject of the experiment. I find this a critical way of monitoring the perfection of one's techniques. Dr. Bazett, who was trained by J.B.S. Haldane, transmitted this to me unequivocally. Until one is willing to undergo the experiments oneself, one must not perform them on other humans. I will oppose those who use patients as experimental subjects before they use themselves. A doctor should never give a drug to a patient until he has tried it himself. A doctor should not insert brain electrodes in patients until he is willing to insert them in his own brain."

Robert: "These are admirable ethics; however, doctors are up against diseases which they do not have. Do you mean that you would subject yourself to, say, an infectious disease agent, the way Walter Reed did with yellow fever?"

John: "Well, I'm not particularly interested in infectious diseases, but I do feel that work on the brain should be done upon intact human subjects who give informed consent. By this, I mean consent based on an understanding of all the dangers, all the procedures, all the techniques to be used."

Robert: "What about brain tumors? Would you induce a brain

tumor in yourself in order to work out techniques of curing it?"

John: "If I were interested in brain tumors, I might try to induce one in myself to find out what patients go through. However, I would not be the same person after the brain tumor that I was before. It is very hard in this area of research to maintain scientific objectivity in regard to something which hits the organ of one's own existence, the brain. No, this is why neurosurgery is an empirical art.

"I see that work on the brain itself may be an area in which the Bazett–Haldane criterion of self as first subject may not apply. According to the basic tenets of my mission, of course, this is a cop-out."

Robert: "Is it a cop-out or is it a beginning realization of what you propose to do in your mission?"

John: "Let us visualize the case in which, by some as yet unknown technique, one is able to control the electrical activity throughout a human brain. By another technique one can pick up the electrical activity of the undamaged human brain in all its myriad parts. Once one has determined on experimental animals that this is a safe technique, then the first subject must be oneself. Experimental controls with animals is a first step; the second step is the use of one's own brain and one's own mind."

Robert: "Let us pursue your fantasy a little further. Let us say that you find the two techniques necessary for exteriorizing the electrical activity of the brain and controlling the electrical activity in an intact, uninjured brain. In both cases we have perfected methods which have no damaging effects on the brain's structure. Let us say that you accomplish this scientific miracle. Your experiments to date do not say that it is even probable, much less possible; however, ignoring that, what are the consequences to you of such a technique?"

John: "Well, such a machine applied to myself would give me an answer to the basic questions, Are the activities of the brain generating the mind or is there something greater than brain activity generating my own personal consciousness? Is the unconscious mind postulated by you inherent in the activities of the brain? Is there something greater than the brain controlling our consciousness? Are we connected to one another and to Beings greater than ourselves by as yet unknown means? Is the brain a leaky container for the mind or is it a valve for a universal consciousness?"

Robert: "Have you thought about the social and political repercussions which might result from such a machine? Let us imagine its impact on the present human society. Let us say that you are able to find the means, that you are able to carry it out. Under whose auspices would this be done? Is there any organization in government, in universities, or anywhere else where such a machine, once developed, would not be misused in the service of current political powers?"

John: "So far I have found no organization to support such work; it is too far out, too fantastic in terms of the human consensus scientific realities. But let us say that somehow or other a private foundation is found. I would not try to do it within government; obviously there are too many conflicting interests. It would inevitably be misused in the hands of the power groups within government. Even modern corporations would misuse such an instrument. I am beginning to see what you mean, I think. You mean that this machine should not even be researched, should not even be constructed, should not even be proposed?"

Robert: "You are projecting those meanings, those judgments, upon me. This is your analysis, not mine. Those are your conclusions, not mine. However, I am glad to see that you are beginning to think in terms of the current level of evolution of Man's institutions. There may possibly be no way of doing what you want to do realistically with the world of Man organized as it is."

John: "As usual you've opened up my mind to possibilities which, in my narrow-visioned way of thinking, were not given their full weight, perhaps not conceived. I am beginning to feel that the analysis is nearing its completion in the sense that I won't need you to listen to my thoughts anymore. I have given you my major secret; you have pried it open and exposed some of its roots, some of its ambitions, and some of its utility."

Robert: "There will come a day in the not-too-distant future

when I believe you will be able to function on your own without
my help. When that day comes, however, I wish you to feel free
to come back and talk over your inner problems with me."

John: "OK. I will feel free to do that. Amazingly enough, I
do not have any negative feelings for you as a consequence of
your objective analysis of my 'secret mission'; in fact, I have
gained a new access to good feelings about the mission. I am glad
it is no longer a secret within me alone. I have shared it with
you, and now maybe I can be more realistic in carrying out my
programs in the future."

Soon after this session John completed his analysis insofar as it
could be done with Robert. He came to the decision to move to
Washington to work at the National Institutes of Health under Dr.
Seymor Kety, who promised him space in two institutes, one devoted to the brain (the National Institute of Neurological Diseases
and Blindness) and one to the mind (the National Institute of Mental Health). He saw that, through his old friend Kety, the possibility existed of pursuing his brain/mind program. He realized that
he must keep one foot in those medical sciences devoted to the
brain and the other in those devoted to the mind until he could see
a secure way of combining the two aspects of research. In the
program he outlined to Robert, he realized that he must continue
to pursue his researches in a dichotomous way, split (as was his
society) between the two views of Man as possessor of a brain and
of a mind, separated according to the limited techniques of that
time. He decided to move to Bethesda and work out the means of
safely doing what he wished to do, first upon animals and then
upon himself. He hoped that his mission had become more realistic and that eventually he could accomplish his aims of developing
a method of recording/controlling brain/mind.

The Lilly Wave

I visualize ten thousand, a hundred thousand, a million electrodes inserted through my skull into my own brain, hooked up to an adequate recording system. —
pages 79–80

Until one is willing to undergo the experiments one-self, one must not perform them on other humans . . . A doctor should not insert brain electrodes in patients until he is willing to insert them into his own brain. —
page 83

Once one has determined on experimental animals that this is a safe technique, then the first subject must be oneself. Experimental controls with animals is a first step; the second step is the use of one's own brain and one's own mind. — page 84

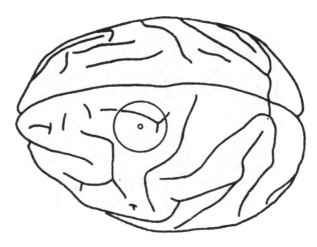

Brain of a monkey. The circle (12 mm. in diameter) shows the location of the trephine hole cut to hold the implanted electrode. The 1 mm. diameter pore was at the locus represented by the center of this circle. "Threshold movements produced by excitation of cerebral cortex and efferent fibers with some parametric regions of rectangular current pulses on cats and monkeys," *J Neurophysiology*, 1952 15:319–341.

Electrical Stimulation of the Brain

Our goal has been to find an electric current waveform with which animals could be stimulated through implanted electrodes for several hours per day for several months without causing irreversible changes in threshold by the passage of current through the tissue.

Many waveforms, including 60-cps. sine-wave current can apparently be used safely for these limited schedules of stimulation. In our experience they cannot be used for the intensive, long-term schedule of chronic stimulation.

Electric current passed through the brain can cause at least two distinct types of injury: thermal and electrolytic.

The technical problem in chronic brain stimulation is to stay above the excitatory threshold and below the injury threshold in the neuronal system under consideration. This result can be achieved most easily by the proper choice of waveforms and their time courses; and less easily by the choice of the range of repetition frequencies and train durations.

— J. C. Lilly, J. R. Hughes, E. C. Alvoro, Jr. and T. W. Galkin, "Brief, non-injurious waveform for stimulating the brain," *Science* 1955 121: 468–469.

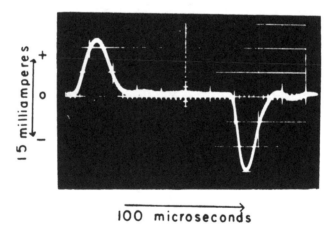

Waveform of stimulating current: pulse pairs of current resulting from quasi-differentiation, with passive electrical elements, of a rectangular pulse. Measured at 2 percent of the peak, the duration of the positive pulse (upward) is 34 µsec., and the duration of the negative pulse (downward) is 28 µsec.

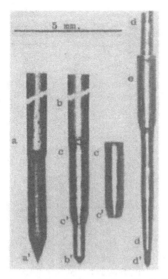

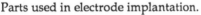

Parts used in electrode implantation.

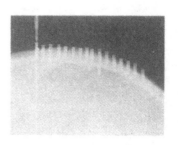

X-ray photograph of monkey's skull (No. 230857, Horatio) containing 20 sleeves and 1 electrode; 16 sleeves are in the midplane and 2 are on each side, 10 mm. lateral to the midplane. J. C. Lilly, "Electrode and Cannulae Implantation in the Brain by a Simple Percutaneous Method," *Science* 127: 1181–1182, 1958.

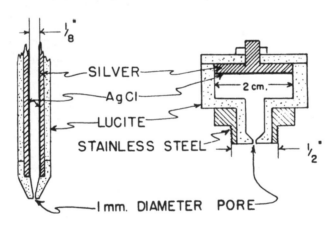

CORTICAL ELECTRODES

Reversible Electrodes for Cortical Stimulation. Both electrodes have a large area of AgCl in contact with the electrolyte bridge to the pial surface. The one on the left (number 1) was built for and used with a Horsley-Clarke instrument for fixation. The one on the right (number 2) is designed for implantation in the skull. "Threshold movements produced by excitation of cerebral cortex and efferent fibers with some parametric regions of rectangular current pulses on cats and monkeys," *J Neurophysiology*, 1952 15:319–341.

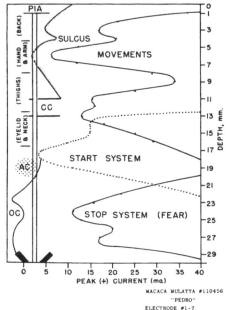

THRESHOLD AND DEPTH ALONG
A SINGLE ELECTRODE TRACK

MACACA MULATTA #110456
"PEDRO"
ELECTRODE #1-7

CC: corpus callosum   OC: optic chiasma
AC: anterior commissure

Thresholds along a single electrode track at 1 mm. intervals. The cortical thresholds show the effect of distance of the electrode from the cortex. The start effect was accompanied by hallucinatory or searching behavior. At levels of current greater than these thresholds in the stop region, a fear syndrome is seen. The stalk of the pituitary is shown at the bottom of the track between about 29 and 35 mm. down. The section is rotated 90° from frontal to sagittal just below the corpus callosum. The track was in the midsagittal plane and went down the stalk of the pituitary. Below 29 mm. only peripheral nerve responses were detected. J. C. Lilly, "Learning Motivated by Subcortical Stimulation: The Start and Stop Patterns of Behavior," reprinted from *Reticular Formation of the Brain,* © Henry Ford Hospital, Detroit, Michigan.

	Kind of Zone in Central Nervous System	
	Start	*Stop*
A. Synonyms	(1) Rewarding (2) Positively reinforcing (3) Positive (4) Enamoring	(1) Punishing (2) Negatively reinforcing (3) Negative (4) Alienating
B. Stimulated behavior, without training, clinical syndromes	Contentment, increased interest, reduction of anxiety, improved cooperation with observer, improved appetite, "want-more" types of reaction	At least 3 related but distinct syndromes: I. Localized pain signs II. Fear to panic III. Malaise to shocklike state In general, escape and avoidance behavior
C. Stimulated learned behavior	In general, start-the-stimulus reactions and this kind of capture pattern	In general, stop-the-stimulus reactions and this kind of capture pattern, and loss of ability to operate a release at high current levels
D. Size of CNS Regions	Large	Much smaller

Control of the Brain
and the Covert
Intelligence Services

⑨

In 1953, when John moved to the National Institutes of Health at Bethesda, Maryland, outside Washington, he took to Bethesda his new machine, which allowed the visualization of the activity of the brain in a television-like display. He had already demonstrated that the electrical activity on the surface of the brain moved in very particular ways across the cerebral cortex in rabbits, cats, and monkeys. In his new laboratory he worked on another means of conducting this electrical activity from inside the brain to the outside apparatus.

At the University of Pennsylvania he had found an electrical wave form which could be used to stimulate the brain without injuring the stimulated nerve cells. He had demonstrated that the previous wave forms used in neurophysiology and in neurosurgery injured the neurons when unidirectional current passed through the brain. He developed a new electrical wave form to balance the current, first in one direction and then, after a brief interval, in the other. Thus ions moving in the neurons would first be pushed one way and then quickly the other way, stimulating the neurons and leaving the ions in their former positions within the neurons. This new wave form was called a balanced bidirectional pulse pair. Microscopic studies of brains stimulated with this balanced pulse pair showed that there was no injury of the neuronal networks from this kind of stimulation. At NIH he continued this work.

In order not to injure the brain substance itself, six hundred and ten electrodes were placed on the cerebral cortex of one macaque monkey. Being able to stimulate the brain through these electrodes without putting the monkey under anesthesia showed that the whole monkey brain was sensory and motor. Every part of the cerebral cortex of the monkey gave responses in some muscle group in the periphery appropriate to that area of cortex.

John's extensive studies on the cerebral cortex were then amplified by stimulation of deeper brain structures. Previous techniques of stimulating deeper structures involved removing fairly large areas of skull, placing the electrodes, and closing the holes with plastic. His previous work with these techniques showed him that the brain was injured by its brief swelling through these holes. He decided that removal of the skull was also a damaging factor in the experiments. He utilized a new technique he called the sleeve-guide technique.

With this technique electrodes could be implanted in the brain without using anesthesia. During the process of implantation, there was no more pain to the animal than that of a needle prick in the scalp.

Short lengths of hypodermic needle tubing equal in length to the thickness of the skull were quickly pounded through the scalp into the skull. These stainless steel guides furnished passageways for the insertion of electrodes into the brain to any desired distance and at any desired location from the cortex down to the bottom of the skull.

As many sleeve guides could be implanted as were desired. Because of the small size of the sleeve guides, the scalp quickly recovered from the small hole made in it, and the sleeve guide remained imbedded in the bone for months to years. At any time he desired the investigator could palpate the scalp and find the location of each of the sleeve guides.

Once one was found he inserted a needle through the scalp into the sleeve guide, down through the bone, and penetrated the dura. After withdrawing the needle the investigator placed a small, sharp electrode in the track made by the needle and pressed the electrode (at the end of a long steel tube with a small diameter) through the scalp, through the skull, through the dura, and down into the substance of the brain itself to any desired depth.

It was found that monkeys subjected to this procedure remained in perfectly good health, with no infection and no detectable aftereffects of the operations.

Since the brain itself has no pain fibers, insertion of the electrode deeper into the brain was not discernible by the monkey. Using these electrodes and the safe, "balanced-pair" wave form, systems were found in several monkeys' brains which when stimulated caused pain, fear, anxiety, and anger. These systems were found to exist mainly along the midplane of the brain, deep, near the bottom of the skull.

Surrounding these were matching systems, which caused pleasure, sexual arousal, and highly positive motivations on the part of the monkey.

Those systems which caused pain, fear, anxiety, and anger were called "negatively reinforcing" systems or, in the shorthand of the laboratory, "negative systems." Those which caused pleasure and positive motivations were called "positively reinforcing" or, in the shorthand, "positive systems."

As the electrode was pushed into the brain one millimeter at a time and each site stimulated in turn, a technique was worked out to decide which site gave positive effects and which gave negative effects. Electronic apparatus was arranged which would turn on or turn off an electrical stimulator. The monkey was given two radio frequency switches, one that would turn the stimulation on and one that would turn it off. When the electrode was in a positive region, it was found that the monkey would quickly learn to touch the "on" switch to turn a short pulse-train on in those sites. In negative spots within the brain, when the investigator turned the stimulus on, the monkey quickly learned to stop it with the "off" switch.

The results in both the positive and negative systems were reproducible within the same monkey and between monkeys. By extensive mapping, involving as many as six hundred sites within a given monkey brain, the positive and negative systems were thoroughly worked out, the locations determined satisfactorily, and the relations between the sites in the neuroanatomy of the brain described in detail. The pain systems were mapped, the sexual systems were mapped.

It was found that in male monkeys there were separate systems for erection, for ejaculation, and for orgasm. With an electrode in the separate orgasm system, the monkey would stimulate this region and go through a total orgasm without erection and without ejaculation. Given the apparatus by which he could stimulate himself once every three minutes for twenty-four hours a day, the monkey stimulated the site and had orgasms every three minutes for sixteen hours and then slept eight hours and started again the next day.

In the negative systems it was found that, if the apparatus was set up to automatically stimulate the monkey once every three minutes for twenty-four hours a day, it would shut off the stimulus each time during the twenty-four hours. If this stimulus was continued for too many hours, the monkey became sicker and sicker; finally unable to push the switch, it headed toward its own death. This clinical picture could be changed by allowing the monkey to stimulate the positive systems which completely reversed the "dying" syndrome.

Motion pictures made of these experiments were shown at scientific meetings. Personnel within both institutes saw them, displaying great interest, and demonstrations were made for visiting scientists.

The directors of the two institutes received monthly reports on this brain stimulation work. As the techniques became known in the institutes, the news traveled through the government as well.

One day John was called to the telephone in his laboratory. The director of the National Institute of Mental Health was on the phone. "John, I have a request for you to brief a meeting of the combined intelligence services of the United States government—the FBI, the CIA, Air Force Intelligence, the Office of Naval Intelligence, the National Security Agency, Army Intelligence, and the State Department. They would like you to make a presentation of your brain-electrode techniques for stimulating motivations within the brain."

John: "Bob, this is a very dangerous area and I am very reluctant to do this briefing. What are the conditions under which it is to be done?"

The director: "You have to set up those conditions before you give the briefing. I know your reluctance to work under secret auspices."

John: "It seems to me that this is a very potent method of controlling human motivations, both positively and negatively. I do not want to do such a briefing under security."

The director: "Despite the fact that you are an officer in the Commissioned Officers Corps of the United States Public Health Service, I will not order you to give a secret briefing. Dr. Kety has told me that all the work in his laboratory and in yours is to be open, not under security. I agree with this."

John: "Dr. Antoine Rémond, using our techniques in Paris, has demonstrated that this method of stimulation of the brain can be applied to the human without the help of a neurosurgeon; he is doing it in his office in Paris without neurosurgical supervision. This means that anybody with the proper apparatus can carry this out on a human being covertly, with no external signs that electrodes have been used on that person. I feel that if this technique got into the hands of a secret agency, they would have total control over a human being and be able to change his beliefs extremely quickly, leaving very little evidence of what they had done."

The director: "Since this is so, I would suggest that you do this briefing under special conditions, that it not be secret, that it be open."

John: "I learned in World War Two that the technical secrecy which these agencies use is its own enemy. The criteria which they apply, the 'need to know,' generate controls over scientists that cause their work to deteriorate. The scientists are unable to talk to those who can probably offer the help and ideas they need to complete their research. My impulse is not to give this briefing."

The director: "Well, that's up to you. With adequate publicity and by forcing them to agree to make it open, you can prevent misuse. A representative of the group that wants the briefing will give you a call."

Several days later John received a telephone call from an unknown man within the United States government, who said, "Your director tells me that you are willing to give a briefing on your work with brain electrodes for controlling motivation by brain stimulation. He says that you have certain conditions for this briefing. What are they?"

John: "I will agree to inform members of the intelligence services on the use of brain electrodes and the techniques of inserting these electrodes and stimulating selected sites under the following conditions: First, the fact of this briefing will not be secret or under any form of security. Second, everything I say at the meeting is to be unclassified, open, and repeatable subsequently by me or by others. Third, at any future time I will be free to write about this conference without any restrictions in regard to what others at the conference say or the context which I present to them. Fourth, any motion pictures or other materials which I present to the group are open. They have either been published or will be published in the future in the open literature."

There was a long silence before the man at the other end said, "I will have to refer this to the group that wants the briefing and the man who is organizing the meeting. I will call you back in about ten days."

Ten days later the man called back. "Your conditions are acceptable."

At the appointed time and place John walked into the meeting and looked around the long table. He saw approximately thirty people, half of whom were in various uniforms of the United States government. The man who had telephoned him introduced him to the meeting but did not volunteer to introduce anyone present to him.

John presented the essential work, showing the types of motivational systems, positive and negative, within the brain. He showed how the monkey could be taught rather rapidly to push a switch to stimulate its own positive systems and to push another switch to cut off stimulation in the negative systems. He showed motion pictures of these findings with the monkeys and slides giving the details of the technique. His demonstration took about one hour.

When he finished there was a long silence.

One man in a blue uniform asked the only question: "What are the medical indications for the use of brain electrodes in human patients?"

John: "The only current medical indications are two rather severe diseases: epilepsy which cannot be controlled by chemical means and Parkinson's disease."

The implications of this question, in the setting in which it was asked, were not lost upon John. These techniques could become the most powerful brainwashing methods devised by man.

By this time John had already begun his momentous research in Florida on the brains of dolphins—research that was to have a profound effect on his life.

Following the briefing of the covert intelligence services on the subject of brain-electrode control of behavior, John received a telephone call from a man from the Sandia Corporation in New Mexico.

"I wish to learn the technique of inserting the sleeve guides into the heads of large animals," he said. "On your next experiment with dolphins I would like to come, take motion pictures of the technique, and learn how to do this."

John: "I understand that the Sandia Corporation is doing classified work. I will agree to show you these techniques but the movies that you take, everything you learn, and the fact that I am teaching you these techniques must not be classified."

On John's next trip to Florida the man from the Sandia Corporation photographed the experiments on dolphins. He agreed to send a copy of the film to John and promised that the film itself would not be classified.

After a few weeks, when a copy failed to arrive, John called the Sandia Corporation and was told that the film had been classified top secret

John went to Washington and talked to a friend of his in the Department of Defense Office of Science. He told his friend of the arrangement he had made with the man at Sandia. The Department of Defense man called the Sandia Corporation, talked to the security officer who had done the classification, and asked that the film be declassified and sent to him in Washington.

During his visit to Florida, the man from Sandia did not say how he was going to use the brain electrodes in animals. He stated that he was working under security and could not tell John the end use for the brain-electrode techniques.

After John left the National Institutes of Health and was searching for means to support the dolphin project in the Virgin Islands, he was asked by the Office of Naval Research to present the dolphin work at the Department of Defense in Washington. He gave a talk to the assembled government people from the Navy and other agencies at the Pentagon.

When John finished his talk, the man from the Office of Naval Research who had invited him to speak asked him to stay and hear a talk by someone from the Sandia Corporation about the use of brain electrodes in animals for carrying loads across mountain ranges. John recognized the person's name; it was the man to whom he had taught the brain-electrode technique.

After John had taken his place in the audience, a security officer came into the room and spoke to the man from the Office of Naval Research (ONR), who then asked John to leave the room with him. Out in the hall he said, "There is some problem about your security file. You are not to be allowed to listen to the next presentation."

John, quite frightened and rather angry, said, "Why is that?"

The ONR man said, "I do not know, but you cannot go back in that room."

John immediately went off to see his friend in the Department of Defense Office of Science and asked him to find out what was going on.

He called several people in the Department of Defense and finally told John, "All I can find out is that apparently, there is some problem with your security file having to do with the FBI. I would appreciate your not giving my name when you inquire about this over at FBI."

John said, "OK. I promise not to use your name."

John went outside to a telephone and called a friend of his father's who at that time was the Assistant Secretary of the Treasury. He made an appointment and went to see him.

In his office in Treasury, his father's friend called J. Edgar Hoover, who referred him to an assistant named Richard Krant. He made an appointment for John with Mr. Krant.

Mr. Krant said, "What is the name of your friend in the Department of Defense who told you that this is a problem involving your security file at the FBI?"

John: "I cannot tell you his name until he gives me permission to do so. I promised him that until he could thoroughly investigate this problem within the Department of Defense, I would not reveal his name. At some future time he will relieve me of this promise."

Krant said, "OK. I will be in contact with you. Give me your telephone number and address in Miami."

The next day in Miami two tough-looking FBI agents came into John's house. They sat down on each side of John. In a very threatening tone, they began to quiz him about the name of the man in the Department of Defense who had said the FBI was at fault in this security matter.

John: "Do you mean to say that the FBI is more worried about its own reputation than about the problem of a citizen falsely accused of being a security risk?"

They said, "We have been told to find out the name of this man and to give you no further information."

John, losing his temper: "Get on that telephone and call Krant and make it collect."

One of the agents made the call and John began shouting at Krant into the telephone. "I agreed to tell you the name of that man when he released me from the promise that I made to him. Now please get these two thugs out of here."

One of the agents came to the phone to talk to Krant, and,

angrily, they both left the house. John called his friend, the Assistant Secretary of the Treasury, and told him what had happened.

Within three weeks John received a telephone call from an Assistant Secretary of Defense who had a house in Florida. He asked him to come see him about the security matter.

The Assistant Secretary of Defense said, "The fault in your security file was a clerical error. The clerk had placed the record of a convicted criminal with the same name as yours in your file by mistake. What can we do to give you satisfaction in regard to this error?"

John said, "I would like a letter of apology from the Secretary of Defense. Copies of this letter are then to be placed in all intelligence agency files in Washington, including FBI, Air Force Intelligence, the Office of Naval Intelligence, CIA, and so forth. In that letter I wish it to be made clear that I am not a security risk, that I have never been a security risk. I want it made clear that the mistake was made in the Department of Defense, Security Office."

He said, "I am afraid we are unable to acknowledge that the mistake was ours. We will be glad to write a letter of apology from the Secretary of Defense, but we cannot accept the blame. We will agree to distribute that letter to the various intelligence services for their files."

The letter from the Secretary of Defense arrived the following week. It was a very diplomatic apology, not taking the blame, and the distribution list was quoted at the bottom of the letter.

As *Harper's* magazine later revealed, the man from Sandia Corporation who followed John at the conference sponsored by the Office of Naval Research was indeed the man John had instructed in brain-electrode techniques on dolphins. At that conference he had shown a movie of a mule going across very mountainous, steep slopes, controlled by a sun compass and brain electrodes. The mule's course was maintained in a perfectly straight line irrespective of terrain. The sun compass was hooked to the brain electrodes so that if the mule deviated from his course he was punished, and if he remained on course he was rewarded, by the appropriate electrodes. The mule could also be radio controlled from a distance to change his course, depending on the desires of those directing him.

Since the Sandia Corporation's main mission was the development of small atomic weapons, obviously this was a delivery system for such weapons which was not subject to the usual radar and other means of detection for metallic vehicles and for heat-generating motors. Such a method would be useful in limited warfare in rough terrain and would also be usable by terrorists. Until new means of defense against this kind of attack were devised, it would have decisive surprise value.

The same means applied to human agents could be used to change their belief systems and also to control them at a distance.

Going over these events, John realized the truth of what Robert had told him during his psychoanalysis. As soon as his research began to show results, though still far from his goal of achieving a safe way of influencing minds and brains by electronic methods, an immediate application was made by corporations, the covert intelligence services, and the military. This demonstrated that, in the human social reality of the day, it would be impossible for him to carry out his own "secret" mission. He saw that even if he achieved his goal it would be misused in the service of Man's warfare upon Man. Reluctantly, he decided to abandon the use of these methods and their further development. He felt that Man was not yet ready for this kind of power. He finally realized the political implications of his favorite research projects. He abandoned the use of brain electrodes and electronic means of finding the relation of the mind to the brain. Almost reluctantly, he turned in other directions. He began to concentrate on communicating with dolphins without neurophysiological aids. He pursued research on the effects of isolation using the flotation tank.

Learning How
to Isolate
Brain and Mind
10

As progress developed in neurophysiological research concerning the physical performance of the brain itself, the scientist in the National Institute of Mental Health came upon a dichotomy in the interpretation of the actions of the brain. There were two schools of thought about the origins of conscious activity within the brain.

The first school hypothesized that the brain needed stimulation from external reality to keep its conscious states going. This school maintained that sleep resulted as soon as the brain was freed of external stimulation. When one retired to bed in the darkness of the night and the silence of the bedroom, the brain automatically went to sleep as a result of its release from the necessity of carrying on transactions with the external world.

The second school maintained that the activities of the brain were inherently autorhythmic; in other words, within the brain substance itself were cells that tended to continue their oscillations without the necessity of any external stimuli. According to this interpretation, the origins of consciousness were in the natural rhythms of the brain's cell circuitry itself.

The scientist contemplated the literature, talked to people involved in these two schools of thought, and decided to do some experiments to test the rival hypotheses.

He reviewed what was known about sleep, anesthesia, coma, accidental injury to the brain, and other causes for the apparent cessation of consciousness in human beings.

He looked at the physics and the biophysics of stimulation of the body. He considered what, according to our present scientific consensus, is thought to stimulate a body. He considered the effects of light and its stimulation of the eyes. He considered sound and its stimulation of the ears, touch and pressure and their stimulation of the skin and deep-lying end organs within the body. He took a good look at the effects of gravitation and its determination of body position and motion. He looked at temperature differences, at clothing, at the effects of heat and cold.

He realized that the important experiments must cut down these various forms of stimulation of the body to the minimum possible value, short of cutting nerve fibers going to the brain. He realized the essential feedback relationship between motion of the body and its self-stimulation during motion—the feedback from muscles, joints, bones, and skin.

He then devised a set of standards for isolating the body from all known forms of physical stimulation. His reasoning was somewhat as follows: Sound and light could be eliminated by well-known methods using unlit soundproof chambers. Motion of the body could be eliminated by voluntary effort with relaxation in the horizontal, gravitational, equipotential plane. The usual bedroom situation, lying horizontal on a bed in a dark, silent room, satisfied these criteria of isolation.

The remaining sources of stimulation, however, were more difficult to remove. When one is lying on a bed, the antigravity pressures on the surface of the body against the bed generate a stagnation of the blood flow to the skin and muscles on the underside of the body. This leads to stimulation which causes the body to turn during the night to relieve the slow flowing of the blood in those tissues under pressure on the bed.

The second group of stimuli which were not satisfied by the bed in the dark, silent room were those of temperature differences—a flow of air as a result of the convection currents over the body cools those parts exposed to the air. Those parts not exposed to air remain warm; in fact, they may become too warm as a result of the local metabolism of the tissues not exposed to the air. The tactile stimulation of the body against the bed and of the coverings over the body would also have to be eliminated.

After a long and protracted study of these various sources of stimulation, the scientist received an inspiration to attempt the use of water flotation. Water would have the advantage of supporting the body without stimulating the skin if there were no motion of the body itself, or of the water relative to the body. If the water temperature were adjusted to soak up the heat given off by the metabolism of the body and the brain, the temperature problem could be solved. Changes in temperature along the body axis and over the surface of the skin could be reduced below the level at which they stimulated the end organs of the skin. The gravitational counterforces supporting the body would also be reduced by flotation below the level at which they could be detected by the end organs and the central nervous system of the body itself.

The scientist thus visualized a tank in which the body could be supported in water that would be maintained at the proper temperature to take care of the generation of heat within the body. This tank should be in a soundproof chamber which could be blacked out. He sketched out the necessary apparatus and began to talk to his colleagues in the National Institutes of Health about this proposal. He realized that in order to furnish air to the person submerged in the water, he would need to draw upon the knowledge of respiratory apparatus learned during World War II in his high-altitude research. The person immersed would need a breathing mask which must be as nonstimulating as possible to the head and face. The apparatus would have to furnish air—the oxygen in the air—and get rid of the carbon dioxide given off by the lungs as a result of the metabolism of the body burning the oxygen. He talked the proposal over with his immediate superior, who suggested that he go to the National Institute of Arthritis and Metabolic Diseases where a man had worked with respiratory systems during World War II for the Office of Naval Research.

He arranged for a luncheon with Dr. Heinz Specht in the cafeteria at the Clinical Center, and in the ensuing conversation Dr. Specht said, "By coincidence there is a suitable facility available in our institute which is not being used."

That afternoon they walked over across the campus of the National Institutes of Health to a small, isolated building labeled T-2. Dr. Specht explained that this was a high-altitude chamber in which animals were being subjected to equivalent altitudes in a series of experiments involving the chronic effect of lowered oxygen on their metabolism. In the back of the building they entered a double soundproof chamber through two locked doors. Inside the chamber was a tank constructed during World War II for experiments by the Office of Naval Research on the metabolism of underwater swimmers.

At this time the scientist realized that somehow or other he had been guided by something outside himself, and greater than himself, which he was later to call "coincidence control." The major portion of the expensive apparatus which he needed already had been constructed and was available for his use. The facility was ideally situated.

Within the institutes it was still possible to do research both on and in isolation. There was no interference from high levels of administrative control. The work could be as open-ended as the scientist wished it to be. He could pursue the ideas and the plans he had set up without reference to the local scientific consensus. This situation existed for a period of two years, during which time he was able to develop the method and to obtain the unexpected results of this research.

Since the scientist was a commissioned officer in the Public Health Service, he was on duty twenty-four hours a day, seven days a week, and had the freedom of the facility at any hour of the day or night. He realized that he must do the isolation work alone in order to obtain results uninfluenced by the social realities. To be in solitude for even brief periods of time meant being free of social transactions and the necessities of interaction with other personnel.

For the isolation research he would disappear from the busy

scene of the laboratory in which the neurophysiology work was taking place for periods of an hour or two hours, or he would disappear at night from his home without telling anyone what he was doing or where.

During the first year of the isolation work, the majority of his time was spent in devising satisfactory masks for underwater breathing. He went to the deep-diving research center of the Navy in Washington, accumulated a collection of underwater breathing masks, and tested them out in the tank in building T-2. He devised a breathing system which would not impede the respiration of the subject floating in the water. It was important to devise this system so that the carbon dioxide could be ventilated adequately and oxygen taken in in requisite amounts without any detectable pressure against respiration.

It was found that all the standard masks, after a half hour or so in the water, pressed on the face, creating low levels of pain and furnishing unwanted stimulation. The scientist learned methods of making his own masks and finally devised one made of latex rubber which covered the whole head and sealed around the neck. An inlet breathing tube came in close to the mouth and nose and an outlet breathing tube left from the same space. In the first models of this mask, there were no eye holes; it covered the head and face and was molded to fit the scientist's own head and face comfortably without pressure above the point of discomfort.

The water that flowed through the tank was tap water, the temperature of which was controlled by a photographic darkroom valve which regulated it at 93 degrees Fahrenheit. It was found that the arms and legs of the body suspended in the water tended to sink. They were held up by surgical rubber dam supports which were minimally stimulating to the skin. This development of a satisfactory apparatus for breathing and for suspending the body took about six months.

Toward the end of 1954, proper experiments using this facility were begun.

The scientist had succeeded in reducing all forms of stimulation to the minimum possible value, and had immersed himself in the darkness, the quiet, and the wetness for many hours at a time.

Within the first few hours of exposure with satisfactory apparatus, he found out which school of thought was correct: the theory that the brain contained self-sustaining oscillators and did not need external forms of stimulation to stay conscious had been proven.

Despite the fact that this choice between the two theories was demonstrated after the first few hours, the research continued as additional data emerged that were of far greater interest.

The scientist made his second discovery: this environment furnished the most profound relaxation and rest that he had ever experienced in this life. It was far superior to a bed for purposes of recuperation from the stresses of the day's work. He discovered that two hours in the tank gave him the rest equivalent to eight hours of sleep on a bed. The two hours were not necessarily spent in sleep. He found that there were many, many states of consciousness, of being, between the usual wide-awake consciousness of participating in an external reality and the unconscious state of deep sleep. He found that he could have voluntary control of these states; that he could have, if he wished, waking dreams, hallucinations; total events could take place in the inner realities that were so brilliant and so "real" they could possibly be mistaken for events in the outside world. In this unique environment, freed of the usual sources of stimulation, he discovered that his mind and his central nervous system functioned in ways to which he had not yet accustomed himself.

He became a bit anxious about these tank experiences. He realized that there were apparent presences which were either created in his imagination or programmed into his brain by unknown sources when he was isolated in the tank. He experienced the presence of persons who he knew were at a distance from the facility. He experienced strange and alien presences with whom he had had no known previous experience. At the time his belief system was such that the brain contained the mind; there were no possibilities such as distant sources communicating with him in the tank. He had had no experience in his scientific work which could account for the experiences that he underwent.

He very cautiously talked to his psychiatric colleagues in the National Institute of Mental Health. He did not tell them of his fears or of his nonconsensus reality experiences. He emphasized the deep relaxation and the benefits derived from his experiences in the tank. Two of the psychiatric researchers decided to try the tank.

The first went in for two hours and came out and reported that nothing had happened. He did not return to the work.

The second did a series of experiments on himself and became hyperenthusiastic about the method, and when he left the National Institute of Mental Health he set up his own tank.

The director of the National Institute of Mental Health Psychiatric Research Division asked the scientist to present a paper at a research conference of the American Psychiatric Association on the results of the use of the tank. About this time the scientist had started to review the literature on solitude, isolation, and confinement. He read everything available on single-handed sailings across oceans, on solitudinous experiences in the polar night; he studied the solitary confinement reports on prisoners of war and prisoners in prisons in the United States.

He found that in none of these situations was the physical isolation as complete as it was in the tank. Most of these experiences had various degrees of stress associated with them due to protracted periods of solitude and of exposure to physical and social extremes. Few of the results were pertinent to his immediate findings except that a person isolated from his own society was likely to know the same freedom and the same variation from the usual experiences of the consensus reality that the scientist had found himself in the tank.

In his scientific paper he very cautiously reported only a few of his less intense experiences. His caution came from the realization that his colleagues would make judgments about his own mental health if he revealed anything more.

Many hours that he had spent in the tank were devoted to his own self-analysis, a continuation of his psychoanalytic work with Robert from 1949 to 1953. The tank experiences gave him new access to bodily pleasure which he found difficult to integrate with his rather, as Robert had called it, Calvinistic conscience. His conflicts with sexual expression, sexual transactions, took up a good deal of his time. The resting body accumulated positive energies that were expressible sexually to an almost intolerable level. He began to recognize the intrinsic nature of sexual drives. His parallel studies in neurophysiology revealed the sources of the sexual energy within the central nervous system. He began to see that these sources existed in himself, in his own brain. He began to see also that the negative energies of anger, of pain, of guilt, were also resident in his own brain, even as he had found them in the monkey brain.

Through such experiences in the tank he began to realize that the cerebral cortex of the human brain is so large that it can suppress, or repress, and hold in abeyance, expression of both the positive, sexual, loving energies and of the negative punishing energies present in the lower circuits. He realized that one does not necessarily have to be a victim of these lower systems, that they can be sublimated or controlled in many more ways than those open to the monkey. He began to see, through his own experiences and the findings of the laboratory, that the huge cerebral cortex of the human was capable of generating any number of alternative courses for the use of these primitive energies.

As he developed the method of tank isolation and began to speak of these results with his colleagues, he discovered a good deal of apparent interest in using the tank method for purposes other than self-analysis and self-exploration. A large section of the psychiatric research division at that time was devoted to research on lysergic acid diethylamide tartrate (LSD-25). The scientist's colleagues in this research suggested that he use LSD in the tank. He refused, saying that he wanted to develop a physiological and psychological baseline of research without using chemical agents first. At that time he did not even contemplate using LSD in the tank; the complexities of the research were not yet sufficiently worked out in his mind to do what he then called "contaminating the results with drug studies." He stayed in contact with the LSD research through his colleagues, however, and attended many seminars on the subject within the institute.

He was bothered by the basic belief systems of psychiatric research. He felt that they were not scientifically based, that they were generated in psychology, psychiatry, and psychoanalysis, none of which were open-ended but tended rather to be closed systems of reasoning, loosely based on Freud's theory of the unconscious mind. In his own analysis he had found that these theories were extremely inadequate as models which would work in the isolated mind.

He had found in his tank experiences that far more was happening than could be accounted for in such theories. The psychiatric profession was not utilizing the full potential of the mind and the brain. Their model was limiting, limited, closed. The explanations for the effects of LSD on the human mind were also limited and closed. LSD was called psychotomimetic; in other words, as if inducing psychosis in an apparently normal person. The belief of the head of the Psychiatric Research Division was that each person had implicitly in him or her a potential psychosis which could be elicited by LSD or by sufficient stress. In other words, self-induced and self-controlled states were not possible once a "psychosis" started. Being exposed to this kind of reasoning, day in and day out, at the National Institute of Mental Health made the scientist very cautious about giving any further information on what was happening to him in the tank. Not until he left the National Institutes of Health and established his own institute in the Virgin Islands was he willing to explore further these deep sources within the human mind which he had detected.

As a by-product of his research in the tank and his work on the brains of the monkeys, he began to consider doing research on animals that had brains equal to or larger than that of man. At the time, he felt that it was important to find out if the large cerebral cortex of a mammal other than man could control the lower systems of positive and negative energies which he had found in monkeys.

In conversations at an international congress of physiology with physiologists about animals with brains larger than that of man, the dolphin was mentioned. Deciding to pursue the matter a year after he had started the tank work, he found that there were dolphins available for research in Florida.

With eight other colleagues from various universities he

went to Florida to investigate the possibility of doing research on the brains of dolphins. Later he went to Florida alone and did crucial experiments on dolphins using brain electrodes. He discovered that they did have control over the stimulation of these primitive energy systems by means of their large cerebral cortex. A dolphin stimulated in the negative regions would not express rage but would shake all over and attempt to terminate such stimulation without panicking. The monkeys had panicked; the dolphins did not. He found the dolphins would control self-stimulation of their positive systems with a great degree of discrimination. He also found that they would use their vocal output to obtain positive-system stimulation, something that the monkeys could not do; thus, the dolphin and the human could control accesses to pleasure vocally. The monkeys' brains were not large enough to do this. Their cerebral cortices had definite limits in the programmatic control of lower systems.

Thus the scientist discovered that the isolation tank could be, in his hands at least, a fertile source of new ideas, new experiences, new integrations. It also brought up past experiences and events which he could not explain with his present limited models of both the human mind and the human brain.

From the tank came the idea of the dolphin research and the stimulus for it. From the tank came a new appreciation of the depth within the human mind unfathomed by previous methods of research. He began to see sources of inspiration, usually neglected, which could be encouraged and realized through the experience of the isolation tank. During this period, though he didn't know it, he was preparing the ground for his own expanded awareness and the achievement of states of being (reminiscent of his early childhood) which he had not yet fully integrated into his personality and was yet to experience totally.

While he was at the National Institutes of Health, the tank work, like the brain-electrode work, also became subject to the politics of the human consensus reality. As the isolation research in the tank became known throughout the government, various individuals called on him to find out about it. Among them were researchers working under the auspices of the Army in regard to

brainwashing of captured prisoners of war. As was the case with
the brain-electrode work, he insisted that such conferences not be
classified, secret or otherwise. He was asked if the isolation tank
could be used to change belief systems of persons under coercion.
Could the isolation tank be used in the service of brainwashing?

John visualized situations in which this method could be
used under coercion; by careful control of the stimulation of isolated persons, their belief systems could be changed in directions
desired by the controlling persons.

John shrank from working out the consequences of this use
of his isolation technique. His experience with the possible misuse of brain electrodes in the service of covert operations had
taught him that this isolation technique could also be misused.
Such approaches to the research further convinced him that he
could not continue to work within governmental agencies.

Conference
of Three Beings

11

One day in 1958 John entered the tank room, put on the mask, and immersed himself in the water for the last time at the National Institutes of Health. He had finally realized that within the government it was impossible to do the research that he wished to do. Inevitably, subtly, those in charge of research for the National Institute of Mental Health were asking to control the isolation-tank work. And those in charge of brain research in the National Institute of Neurological Diseases and Blindness were beginning to exert controls on the work on the brain. In this session in the tank, he planned to review what he had learned over the last five years in regard to research on the brain and the mind and the support of this research.

John went through his now more-or-less standard procedure of relaxing every muscle in his body while floating in the water. He then relaxed his mind and let go of the residues of the day's activities. Quite suddenly he was in a new space, a new domain.

He left his human body behind. He left his human mind behind. He became a point of consciousness, of awareness, in an empty, infinite space filled with light.

Slowly two presences, two Beings, began to approach him from a distance. There was a three-way exchange of direct thought, of direct meaning, of direct feeling with no words.

Later he was to write up the experience as if words were used, as if the two Beings had spoken to him in English, as if he had become the Third Being.

The conference of the three Beings was taking place in a dimensionless space, the spaceless set of dimensions somewhere near the third planet of a small solar system dominated by a type-G star. The organization which they represented he would later call the Earth Coincidence Control Office (ECCO).

The First Being speaks: "We are meeting at this particular space-time juncture in order to review the evolution of a vehicle that we control on the planet Earth. He is at another transition point in his training. We need to review what he has done, what he is thinking, what his motivations are. We must determine what the future of his mission can be within the evolutionary speed limit allowed the humans on that planet."

Second Being: "You, the First Being, and I, the Second Being, have been controlling the coordination of coincidences of this agent on Earth. I feel it is important that we state all of this very clearly for the benefit of the Third Being, who has been responsible for that human agent. It is important that he not exceed the evolutionary speed limit at this particular time; however, we realize that there is a certain discrepancy existing among the humans, that their evolution is proceeding extremely rapidly in certain areas and is going backward in others. It is the purpose of this conference among the three of us to make sure that the Third Being controls him so that he stays within certain well-defined limits and avoids the kinds of catastrophes certain other agents of ours have experienced on that planet. Let us listen to the report of the Being who has been in charge of the vehicle on the planet."

Third Being: "Currently my agent is in a quandary. I need this conference to know in what direction he is to move next. The vehicle that he inhabits is now in a deep trance state and is willing to share with us the sources of this quandary.

"As you both know he has a carefully constructed cover story in which he has invested a good deal of time, effort, and training. All three of us are well acquainted with the rather arduous steps that he has been taking in his human form. There have been many times when he has lost contact with me, has repressed his knowledge of me, and has had to be guided through his unconscious mind. There were times when he had too much knowledge of me, necessitating repression so that he could continue to function as an acceptable human being in the society in which he lives. His main worry at times was being made *persona non grata* by his fellow humans in various fields of endeavor.

"He went through a process which humans call psychoanalysis. Psychoanalysis to us is a means of educating humans in how to remain human, at the same time keeping sufficient independence of that state so as to be aware of our existence. Psychoanalysis also furnishes the human agents with the current rationalizations and basic assumptions upon which humans operate. It helps them develop their cover story so that they will not reveal our existence or our influence. It allows them to reconstruct their past history and understand it in terms of the present human society.

"In the area of self-awareness, his awareness of me, of his deeper self, our agent is on the threshold of recognizing us and our influence on him.

"In his brain research he has discovered the difference between small brains (monkeys) and large brains (human and dolphin). He has realized that in order to do research on the brain and the mind, he must work in an institution which he himself controls, insofar as is possible in the current human reality. He sees that human society interacts with him in a way that allows certain areas of research and not others. He realizes that the laboratory work in relation to the isolation tank is difficult to support. He has learned that the brain work can be supported openly and that the tank work must be done covertly. As long as the tank work was done in solitude, he was unimpeded in the directions in which he could carry it.

"He is becoming aware of the political and social realities of what he is doing. His mission to thoroughly investigate the brain with thousands of electrodes and with feedback between his own brain and its recorded activity has implications which he knows he cannot reveal at the present state of development of the human species. His analyst taught him to look more critically at this particular aspect of his mission. His work for the last five years has been in the direction of perfecting electrodes so that he can use them safely in his own brain. He found on monkeys and dolphins that this was not a safe procedure. The brain sections revealed damage along every electrode track to the extent that he would not wish to insert electrodes into himself or another human.

"From his interactions with the government agency in which he has worked, he finally realizes that the isolation tank gives him more information in more dimensions than can be absorbed by those in control.

"He now wishes to abandon the study of electrodes. They are too damaging to brains. He wants to pursue other methods, not including brain electrodes or damage to either his own brain or monkey or dolphin brains.

"His work to date with dolphins has convinced him that they are quite as intelligent, quite as ethical, quite as sentient as humans.

"He realizes that his first marriage is to a person with whom he cannot share any important considerations. He wishes to obtain a divorce and hopes that he can find a dyadic partner who will share his particular aims and his particular mission."

Second Being: "What are his basic beliefs about the existence of the Third Being and about us?"

Third Being: "He is oscillating between two belief systems. In the first he believes that the mind is the computational software of the brain, that the brain evolved on the planet Earth from the forebears of Man and generated Man's consciousness. In the second system, he believes in us. This belief is contaminated by his childhood faith in the soul. He has yet to develop a pure, integrated view of the mind as an entity not contained in the brain. He has yet to give up the view that the brain contains a computed mind plus access to us through means at present unknown on the planet Earth. When he is in the isolation tank for a sufficient period of time, the second belief system begins to take over. When he is in the laboratory or dealing with the realities of the support of his research, the contained-mind belief system takes over."

Second Being: "I would like to suggest that we arrange for his education in more profound ways. He still needs to penetrate into his own mind deeply in the areas of interest to us."

Third Being: "He is making progress in that direction. Currently his plans are to establish his own laboratory in an isolated location in the Virgin Islands in the Caribbean Sea for research with intact dolphins."

First Being: "When he leaves the government and goes to the Virgin Islands, I suggest that we control the coincidences in the direction of encouraging the dolphin research. He has much to learn about the Beings known as dolphins. We must also control the coincidences in regard to his seeking a female partner for a dyad. He has much to learn here, not all of which can be taught him without some pain. He is not yet in sympathy with the female mind among humans on his planet. He projects too much of himself, not realizing that there are two universes of humans, male and female. Nor is he aware that in the human social reality there are many substructures of the male-female relationship. In his first marriage he was more or less detached. He must now learn to attack negatively to find the parameters of balance in the dyadic domain."

Second Being: "It is felt that coincidences must be regulated to help him continue the isolation work under better circumstances. We should also arrange for him to use LSD-25 in the tank."

Third Being: "We must continue our cover of our existence in his mind: if he is too aware of us at his current stage of training, he will be unable to operate in the human realities. I suggest that we temporarily cut off his awareness of us until later, when he is better prepared to deal with our existence."

First Being: "Let us adjourn this conference and meet at some future time in regard to our business with this agent."

Simulation
and Experience

12

Floating in the tank John slowly came back to his body, to the tank, to the human reality. His exuberance faded and his memory of the conference of the three Beings faded rapidly. His realization that he was still in a government installation inhibited further thought along these lines. He resolved to leave the government to gain more freedom to do what he had to do. Gradually the sense of his human mission reasserted itself and he became once again the scientist.

His protective skepticism also reasserted itself. He said to himself, "This must be another of my 'imaginary' simulations of reality."

The thought occurred to him, I'd best go write this up quickly before the human consensus reality takes over completely.

After showering and putting on human clothes once more, he sat in his office and considered how much of the experience he should write down. What would be the reaction of an eventual audience to this report of his experience? He considered his simulation of his public image, his simulation of the consensus reality, his simulation of "proper science." He weighed the repercussions from the consensus reality of other humans that might affect publication of a book containing such a report. He realized the very narrow framework in which the book must be written in order to be published, bought, read, and understood by a sufficient number of people to carry out the necessities of the mission assigned to him.

As he sat in his office, the first gray light of dawn was appearing. As he considered what to write, the sun came up and shone upon his back. He felt his body, he felt the reality of his office, he felt the reality of writing down what had happened to him. His biocomputer began to generate objections to writing this down. He thought, Was it all merely a dream, an imaginary situation, a science-fiction script?

He thought many alternatives out, considering them all, and decided to write, as best he could, the inner reality as an actual happening insofar as he was allowed to report it. Then he reviewed his theory of the contained versus the uncontained mind.

The mind contained within the brain is the result of the evolutionary process occurring on this planet. The proper atoms accumulated in this portion of the universe at the proper distance from the sun, at the proper temperature. The coalescence of the atoms into a planet, into an atmosphere, into water upon the surface of that planet formed the seas and oceans. The storms over the ocean produced lightning, streaking through the atmosphere and forming nitrogen compounds which fell into the ocean; there combining with the carbon of carbon dioxide, the atoms coalesced into long chains of peptides and eventually proteins. The peptides and the proteins formed small balls of peculiar construction. These balls floated in the bosom of the primordial soup of the seas. The balls coalesced, joined one another, incorporated a new structure. The new structure became the primordial viruses; the viruses joined, became the first bacteria. The bacteria accumulated further layers and became the first protozoan cells. The protozoans joined in colonies, forming the first sponges and the first coral. Evolution proceeded over millions of years, resulting in the worms, the starfish, the tunicates.

Within these organisms new cells developed, specialized for conduction of nervous impulses. The prototype of the nervous system began to evolve in the jellyfish, the worms, the starfish, the tunicates. The fish evolved, the nervous system moved toward the head end of the new organisms. The distance receptors, the eyes, the lateral line organs all became oriented forward in the direction of motion. The lungfish evolved, climbed out on the land, and looked at the new territory. The amphibians evolved in the sea, climbed out on the land, became adapted to a combined dry-land–water existence. The reptiles evolved from the amphibians. Some reptiles returned to the sea and slowly but surely evolved into huge forms which gradually became the primitive dolphins and whales. The nervous system grew and grew and, in the whales, evolved to the size of the humans to come fifty million years later.

The reptiles on the land gave rise to the first landborne mammals. The mammals climbed in the trees and grew. Their brains grew larger and larger, and finally they evolved into the monkeys, the apes, the human species in the prototypic form. The prototype humans evolved further, increasing their brain mass. Meanwhile, in the sea, the dolphins and whales evolved larger nervous systems that equaled the present human one thirty million years ago.

In each of the large evolved animals, including the human, were brains now capable of new choices, of new directions, of new control over the environment and over the self. The contained mind, as we know it today, evolved to its present complex distribution.

As Man became aware of his own awareness, as Man became aware of his own brain, as Man became aware that he was interdependent with other men, as Man built, created, and lived in his human consensus reality, he lost contact with his planet. He developed delusions of grandeur in which he was the preeminent species. He gave himself a special creation and a special evolution separate from that of the planet and from the other creatures created upon it by the evolutionary process.

Man created a special origin for himself. He represented his consciousness as a "soul." The soul was a divine portion of God, the God that he had created in his own contained mind. He assumed that his mind was part of a Universal Mind, uncontained in any brain. He peopled his external reality with simulations of the operations of minds greater than his. He worshiped these minds. He organized churches. He wrote books. He wrote handbooks of God, called the Bible, the Koran, the Sutras, the Vedas, the Upanishads.

Various men and women went into solitude, into hermitages, into remote cells, into caves, into the deserts and experienced realities which they called spiritual. Their minds contained in their brains generated experiences beyond the planet, beyond their human form, beyond their understanding.

Slowly Man began to investigate his own material nature. He studied his brain. He studied damaged human brains. He studied the results of damage to brains. He explored the inner realities of those with intact brains, those with small brains, those with large brains. He traced his own genetic code, his inheritance carried forward in his three-billion-year evolution from primordial matter. He studied the complex assemblages of that matter. He investigated the primitive particles giving rise to that matter. He studied the molecular configurations present in living organisms and in his own body and brain.

Man's evolution began to be what Man conceived it to be: what he thought his fellowman was, his human organizations were, his speech, writing, computers, and the vast diversity of beliefs about self were. Man warred on Man. Man killed Man. Man killed other organisms on the planet by the hundreds of thousands and the millions. Man's own thoughts of himself were as a divinely special creature, evolved and designed to exploit the planet for his own survival and economic gain. His laws were designed to regulate his behavior with regard to other humans and to control the other organisms as if they were a part of his own divinely inherited property. He domesticated many animals; those who could not be domesticated were killed.

The extinction of species proceeded rapidly. Finally some men became aware that if this course continued, the planet would be bereft of many species, many more than those that had already become extinct through Man's efforts. The whales and dolphins of the sea were being killed in the name of the economic religion of

Man. His conceit that other species were an economic resource for his exploitation was killing off huge cultures, huge histories of which Man was not yet aware. Man's development of speech and man-to-man communication isolated him from the possibilities of communication with other brains as large as his and larger. The whale and dolphin cultures were being decimated, their histories stored in their huge brains decimated, as the oldest and largest of the whales were selected for slaughter.

As the scientist reviewed this history he felt as if he were in a human trap, constructed over the millennia.

He thought, "I am merely one of billions of humans. How can I possibly influence the evolution of this planet? Belief is fighting belief in the human reality. Is what I believe to be true, true, any more than what others believe to be true? Is my ignorance any smaller than the ignorance of others? My knowledge feels so small! I want to enlarge it. I am enlarging it, and a temporary set of beliefs to enlarge my knowledge is necessary. Those beliefs can be changed as the knowledge accumulates. What are the paths to knowledge? We as humans need to communicate with others than ourselves to escape this trap of the closed system of our own communication: this trap of our own competitions, of our human-to-human warfare, of our human-to-human dogmatic beliefs, of our human-to-human lethality, of our human-to-other-species lethality. We must control our killing, our devastation, and develop understanding. How can I best effect this?

"For the acquisition of new knowledge one needs the cooperation of selected humans, self-selected humans. Humans with discipline, humans with knowledge of science, humans with flexible new belief systems beyond those which have kept us in our present prison of belief. Knowledge of brains. The alternatives open to large brains, especially those beyond our own. If the mind is contained in the brain, then the size of that mind is a definite function of the size of that brain. The number of alternatives open to a mind is a function of how large the brain is in which it is contained. The uncontained mind belief, if true, then says that what the mind has available to it through the vehicle in which it resides is a function of the size of the brain available to that mind. The material valve for Universal Mind regulates the amount of revelation, the amount of knowledge which can be squeezed through that valve.

"I swing from contained to uncontained mind and back to contained mind. I swing from belief in the three Beings to the simulation of three Beings as a convenient method of thought to free up my thinking. Belief versus experience. Is belief any truer than experience? Is direct inner experience truer than direct outer experience? Are the consensus beliefs applied to belief itself and to experience itself, inner and outer, valid?

"Who are the whales, dolphins, and porpoises? Will we ever know? We won't know until we break the communication barrier with them. I hope we are capable of doing that, and I hope to devote a good deal of energy, time, and money to that program before it is too late, before there are no Cetacea left to communicate with."

John thought, "Thus, in the sermon for today, if my mind is contained within my brain, then the three Beings are either leakages of information into a leaky mind, or they are constructed simulations for unknown purposes, created within my own brain from sources not yet available to me consciously. As Freud would say, they are constructions from my unconscious, generated by belief systems put into me as a child.

"If my mind contained in the brain is leaky to sources of information not yet contained within our knowledge, within our science, then there may be other intelligences in this universe with whom we can communicate and do communicate when we are in the proper state of consciousness and proper state of being. If the mind and the brain are leaky, then they are available for networks of communication beyond our present understanding. If the mind is uncontained, then the Beings have an objective, verifiable existence which others can also share. If I believe in the uncontained mind, then the program I have outlined here can be believed by some others, and we can have a parallel programming into his belief system in which the Beings are 'real.'

Thus we construct another simulated reality as if true.

"If the mind in the brain has sources of information beyond the present science, then there is something or someone communicating with us. We arbitrarily assign descriptions of this someone or something which correspond to ourselves. We project our simulations on the information coming in to us from unknown sources.

"If our minds are really uncontained, then there is no science; there is only education of us by the Universal Mind. Our brains are limited valves, cloaking what works on this planet with what is fed to us but does not work on this planet. If we believe that what works is all there is, then we close our minds and contain them in order to do the empirical job of surviving and generate the illusion of a human evolution on this planet.

"Sometimes I believe we are alone in this universe, that we are strictly an accident, that we originated in the primordial soup of the seas, and that matter has its own evolutionary laws which are dimly understood by us, who are a product of that evolution. Maybe Earth is the only incubator of life as we know it in the whole galaxy. If this thought is true, then we merely construct dreams from the noise of our own brains and from the cosmic noise of our radio telescopes. If this thought is real, then we are only projecting that which we create within our biocomputers back onto the universe and back onto our own structure.

"Freud wrote in his discussion of religion, called *The Future of an Illusion,* 'No, science is no illusion. But it is an illusion to suppose that we can get anywhere else what science cannot give us.'

"Is this my true belief? If so, my ignorance is profound. Science has not yet been able to construct models of us, models of the universe, which are satisfactory. My hunger for simulations of reality which work is not satisfied. Science to me is open-ended, not closed, and cannot be closed within my lifetime. The unknown is still with me, my ignorance is still vast. My knowledge is so small, so puny, so weak, can I somehow escape the narcissistic human-centered sources of knowledge and move into new knowledge with other species?

"I always seem to end on this question. It is time to do

something about it in a way that is demonstrable to others with sufficient knowledge to realize new methods. The new methods that we conceive of must be made real in the form of hardware to solve the problems of inter-species communication.

"If I am really being guided by Beings in other dimensions, then here's hoping that they will guide me along this path to solve these problems before I die. If I am projecting these Beings, at least I can use this simulation as an inspiration and as a source of opening my mind to new potentialities within the possible and within the probable, making real what is possible to make real in the future, sharing it with those other humans who are also involved in scientific demonstrations."

Within the next few days John made appointments with his two directors, resigning from the Commissioned Officers' Corps of the United States Public Health Service and from the two institutes.

He left Washington, went to the Virgin Islands, explored possible sites for a laboratory, and lived alone for a period of a year.

During that year he obtained a divorce from his first wife and made the necessary contacts to set up his new laboratory.

The Lilly Tank

We have been seeking answers to the question of what happens to a brain & its contained mind in the relative absence of physical stimulation. In neurophysiology, this is one form of the question: Freed of normal efferent and afferent activities, does the activity of the brain soon become that of coma or sleep, or is there some inherent mechanism which keeps it going, a pacemaker of the "awake" type of activity?

In our experiments, the subject is suspended with the body and all but the top of the head immersed in a tank containing slowly flowing water at 34.5° C. (94.5° F.), wears a blacked-out mask (enclosing the whole head) for breathing, and wears nothing else. The water temperature is such that the subject feels neither hot nor cold. The experience is such that one tactually feels the supports and the mask, but not much else; a large fraction of the usual pressures on the body caused by gravity are lacking. The sound level is low; one hears only one's own breathing and some faint water sounds from the piping; the water-air interface does not transmit airborne sounds very efficiently. It is one of the most even and monotonous environments I have experienced. After the initial training period, no observer is present. Immediately after exposure, the subject writes personal notes on his experience.

In these experiments, the subject always has a full night's rest before entering the tank. Instructions are to inhibit all movement as far as possible. An initial set of training exposures overcomes the fears of the situation itself.

In the tank, the following stages have been experienced: (1) For about the first three-quarters of an hour, the day's residues are predominant. One is aware of the surroundings, recent problems, etc. (2) Gradually, one begins to relax and more or less enjoy the experience. The feeling of being isolated in space and having nothing to do is restful and relaxing at this stage. (3) But slowly, during the next hour, a tension develops which can be called a "stimulus-action" hunger; hidden methods of self-stimulation develop: twitching muscles, slow swimming movements (which cause sensations as the water flows by the skin), stroking one finger with another, etc. If one can inhibit such maneuvers long enough, intense satisfaction is derived from later self-stimulations. (4) If inhibition can win out, the tension may ultimately develop to the point of forcing the subject to leave the tank. (5) Meanwhile, the attention is drawn powerfully to any residual stimulus: the mask, the suspension, each come in for their share of concentration. Such residual stimuli become the whole content of consciousness to an almost unbearable degree. (6) If this stage is passed without leaving the tank, one notices that one's thoughts have shifted from a directed type of thinking about problems to reveries and fantasies of a highly personal and emotionally charged nature. These are too personal to relate publicly, and probably vary greatly from subject to subject. The individual reactions to such fantasy material also probably varies considerably, from complete suppression to relaxing and enjoying them. (7) If the tension and the fantasies are withstood, one may experience the furthest stage which we have yet explored: projection of visual imagery. I have seen this once, after a two-and-one-half hour period. The black curtain in front of the eyes (such as one "sees" in a dark room with eyes closed) gradually opens out into a three-dimensional, dark, empty space in front of the body. This phenomenon captures one's interest immediately, and one waits to find out what comes next. Gradually forms of the type sometimes seen in hypnogogic states appear. In this case, they were small, strangely shaped objects with self-luminous borders. A tunnel whose inside "space" seemed to be emitting a blue light then appeared straight ahead. About this time, this experiment was terminated by a leakage of water into the mask through a faulty connector to the inspiratory tube.

In our experiments, we notice that after immersion the day apparently is started over, i.e. the subject feels as if he has just arisen from bed afresh; this effect persists, and the subject finds he is out of step with the clock for the rest of that day. He also has to readjust to social intercourse in subtle ways. The night of the day of the exposure he finds that his bed exerts great pressure against his body. No bed is as comfortable as floating in water. J. C. Lilly, "Mental Effects of Reduction of Ordinary Levels of Physical Stimuli on Intact, Healthy Persons," *Psychiatric Reports*, 1956.

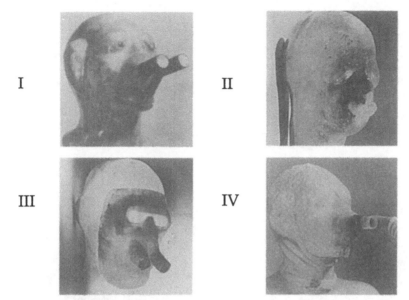

Underwater Mask Manufacture I
Positive plaster cast of one subject's head and neck (J.S.), with spaces over eyes, nose, and mouth, and ears filled in, and (bakelite) tubing in place. Note smooth surface from which later casts in plaster or latex may easily be separated.

Underwater Mask Manufacture II
Dipping form for one subject (J.L.). A plaster bandage cast (negative) is made from the modified reproduction of the head and neck casting (#1) with a separating agent. Inside this negative mold, fiberglass cloth and resin is inserted to construct the positive reproduction shown in this figure. The two white areas between nose and chin are the bumps for positioning the two tubes shown in Figure 3. The two halves of the fiberglass reproduction are assembled and fastened together with fiberglass tape and resin, and finally sanded. A base with a socket for a 75-watt heating lamp is inserted in this model.

Underwater Mask Manufacture III
The dipping form: mask insert in place on the face. (Heating lamp is lighted.) The mask insert is built of fiberglass on the dipping form with latex as the separating agent. The inspiratory and expiratory tubes (stainless steel) and the window (Lucite) are inserted and embedded in the fiberglass at the proper stages; three layers are built up. This form is separately dipped into latex, put on the headform, and the assembly dipped to generate the head mask (three to five layers of latex suffice).

Underwater Mask Manufacture IV
This is a finished mask formed by dipping cast shown in Figure 1, which does not contain a lucite window before the eyes — note breathing tube connected on one side. Note also the loose folds of latex at the neck which seal out liquid without undue constriction of subject's neck. Mask is donned simply by pulling down over head, much as a stocking would be. In place, it fits face and head closely, is smooth, and is completely comfortable, even for long time periods.

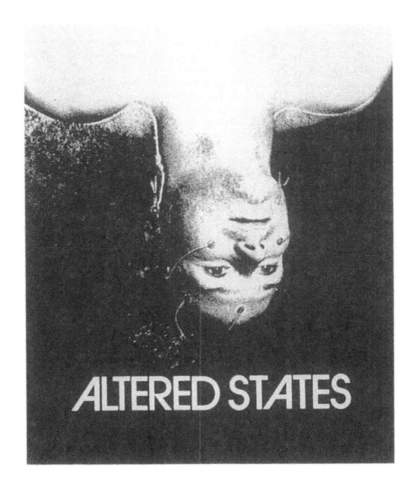

Paddy Chayefsky's movie *Altered States* was based on the work of John Lilly.

By the 1980s, isolation tanks became popularly available from producers, like the samadhi apartment-tank shown here.

At the National Institute for Mental Health, I devised the isolation tank. I made so many discoveries that I didn't dare tell the psychiatric group about it at all because they would've said I was psychotic. I found the isolation tank was a hole in the universe. I gradually began to see through to another reality. It scared me. I didn't know about alternate realities at that time, but I was experiencing them right and left without any LSD.

Various LSD pushers at N.I.M.H. were insisting that I take it but I didn't until ten years later. I finally took it in the isolation tank in St. Thomas in the Virgin Islands with three dolphins there. I took LSD for the first time, in the tank, with three dolphins under it in a sea pool. I was scared shitless. It was 300 micrograms injected intramuscularly. As I climbed over the wall into the saltwater, a memorandum from N.I.M.H. appeared before me — "Never take LSD alone." That's when I learned that fear can propel you in a rocketship to far out places. That first trip was a propulsion into domains and realities that I couldn't even recount when I came back. But I knew that I had expanded way beyond anything I had ever experienced before, and as I was squeezed back into the human frame, I cried.

Transition

13

By a series of coincidences John found the means to buy some land in a suitable location on the island of St. Thomas in the Caribbean Sea where he could build his new laboratory. He invested what money he had available and was able to obtain other funds from the National Science Foundation. The building of the laboratory began.

At the National Institutes of Health, he realized that physiological research would be needed in the new institute in order to support work involving communication with the dolphins as well as the tank work. He would need scientific colleagues to do the physiological work and another location for that work outside the Virgin Islands. He found colleagues to control the physiological work and set up a laboratory in Miami while the laboratory in the Virgin Islands was being built. During the next few years the two laboratories became operational. The brain work (physiology) and the mind work (communication with dolphins and tank research) were to be split between Miami and the Virgin Islands. The two laboratories were far enough apart so that the personnel would not interfere with one another. Those who wished to pursue the physiology work were in Miami; those who wished to pursue the communication work were in the Virgin Islands.

He soon remarried and his first daughter was born in this

marriage. As the years went by, his wife insisted on living in Miami. She became involved in the administration of the Miami laboratory.

The St. Thomas laboratory became devoted to an attempt to teach the dolphins to speak English. By coincidence a talented young woman by the name of Margaret Howe was able to spend full time on this project.

At the Virgin Islands laboratory an isolation tank was built. In Miami a thorough study of the brain of the dolphin was initiated by Dr. Peter Morgane under John's direction.

When John was satisfied that the laboratory in Miami and the dolphin communication research in St. Thomas were functioning satisfactorily, he started a series of isolation tank experiences himself in the Virgin Islands.

Through his former colleagues in the National Institute of Mental Health, he obtained some LSD-25 from the Sandoz Company. At that time he had a fellowship from the National Institutes of Health and was permitted to do the LSD research under their auspices. He disguised his true motivations for using the LSD-25 in a research proposal "to see its effects upon dolphins."

He carried out a series of experiments on six dolphins and found the critical dose for the first behavioral change detectable by human observers; 100 micrograms was a sufficient dose for these effects. It had previously been found that this was approximately the critical dose for behavioral changes in small monkeys and in humans. He concluded that the body weight of the particular subject made no difference in the effects of LSD. Most other chemical agents tried on animals and humans, such as barbiturates, seemed to depend on a body-weight effect. The dilution of the chemical in the blood and other body fluids did not make the same difference with LSD. It looked as if LSD was somehow concentrated by the brain no matter the size of the body over a range from approximately three pounds of body weight (monkey) to four hundred and fifty pounds of body weight (dolphin).

After he found this critical value, he decided to continue the initial work with LSD on himself, alone in the tank. During the following two years he was able to carry out a series of such experiments.

At this time John felt that he had completed certain preliminaries necessary to start the new research with LSD. In 1960 he had written a book at the instigation of Herman Wouk which was published by Doubleday the following year under the title *Man and Dolphin*. From 1961 to 1964 he had obtained support for the laboratories in Miami and in St. Thomas. He had organized these laboratories to be self-sustaining for periods up to five days at a time. Supervising the work in the two laboratories required him to travel from Miami to St. Thomas and back several times a month. He felt that he would be able to do isolation work in the new tank in St. Thomas, using LSD, without interfering with the dolphin research either in St. Thomas or in Miami.

Since leaving the NIH he had not had a tank available and missed the inner disciplines and help of the tank in the problems of external reality. He felt strongly that it was necessary to get back to tank work. He realized that the LSD plus the tank experiments must be done covertly; the existence of this research had to be hidden from his colleagues and from those in support of the dolphin programs. His mission to penetrate ever more deeply into the human mind, using his own as an example of that of his species, drove him on into rather dangerous regions in terms of the social consensus reality and his own reputation and public integrity.

At that time there were rumblings in the medical research circles that those who took LSD were somehow different from those who did not. There was a tendency among the medical researchers to discount the findings of those who had taken it. John realized the various causes for this.

Some researchers had become hyperenthusiastic in an almost unreasonable way about LSD and its effect on their own minds. Others who had taken it were frightened off by their first sessions. Various extreme claims were being made for its therapeutic benefits in certain diseases, including alcoholism and neurosis. Those who had not taken it did not realize its profound effect on the human mind. Some of those who had taken it several times and were trained in its effects upon themselves and patients were claiming benefits which could not be visualized or understood by their opposition. Certain researchers had dropped out of the research field and were publicly proclaiming LSD as some special "sacrament" in new religions for all humans.

Various patients who had taken many therapeutic sessions with LSD were writing books containing accounts of experiences that no rational, consensus-reality-oriented person who had not taken LSD could understand or could keep from fearing.

John's previous work in the tank had convinced him that the range of the human mind was far greater than those who had not taken LSD had ever suspected. He was aware of a scientific puzzle in the effects of LSD on the human mind. He had become intrigued enough to explore these effects directly on himself. The Bazett-Haldane criterion of research on human subjects, "Do on yourself that which you would do on others before you do it on others," was still his scientific ideal. Driven by his curiosity and that ideal, he decided to initiate this series of experiments on himself in the tank in St. Thomas.

He realized the social and personal dangers involved and hence took only a few others, sympathetic to this point of view, into his confidence.

Two Beliefs

14

The human brain, a living computer with unknown properties. A biocomputer. Its properties only dimly understood. Its programs for fear, for anger, for love, for pleasure, for pain built into its structure, its circuitry. Circuits for all these feelings are primitive, inherited in the genetic code. The mind is the software of that biocomputer. The observer located in that brain is the result of the brain's computations. As pain and pleasure are inalterably linked in the biofeedback of the brain with the body, so is the observer computed in the large cerebral cortex.

The observer is a programmer resident in the brain. The programmer is the agent within the brain. The self is the operator contained within the brain in self-reflexive circuitry in the cerebral cortex.

Is the self immersed in the brain anything more than the computed result of the brain's software? Is the mind anything more than the computational activities of that brain?

Before starting the first injection of LSD into his thigh muscles, John wrote this preprogram for the whole series of experiments. These were the questions that he set out to answer insofar as he was able. Nothing in his first two training sessions with LSD supervised by a friend in California had given him the answers to them. His analysis of the results of those two sessions did not give him sufficiently accurate answers. Everything he had learned from them reinforced his belief in the containment of his mind within the computational domains of his brain. Everything that had occurred in those two sessions could be accounted for by the contained-mind hypothesis in which he believed at that time.

In the complete blackness of the isolation room at the edge of the tank he injected 100 micrograms of pure Sandoz LSD-25. He entered the tank and floated at the surface of the 93-degree-Fahrenheit seawater pumped in from the Caribbean and warmed in the tank.

His fear was profound. No one had ever done this experiment before; it had never been performed either on himself or on anyone else. The warnings of the staff at the National Institute of Mental Health entered his consciousness. "No one is to take LSD in solitude without a guide. The staff in the LSD research program have had some rather terrifying experiences when taking it alone. The director of the program forbids solitary administration of this agent in the research staff."

His floating body shook with the fear; it fast became terror and then panic. He was tempted to climb out of the tank but realized that this in itself might have been a dangerous procedure, climbing over the eight-foot-high wall under the influence of LSD. He stayed floating and began his preprogramming once again.

"This fear is being generated by memory. Memory of that memorandum. The fear is coming from the lower circuits within my own brain and is expressing itself in my body. My negative reinforcement circuits are hyperactive. I have sufficient connections with those circuits to attenuate their activity. I can neutralize that fear without suppressing the energy. This is the first experiment that anyone has done of this type. I must summon up my courage and my control in order to come through this intact."

The LSD effect started. The electrical excitations traveling through the body were very familiar. He exerted the discipline learned in previous tank work and allowed the body to relax in the presence of the "electrical storms" moving through it. The darkness, the silence, the wetness, and the warmth disappeared. The external reality of the tank disappeared.

His body disappeared: his conscious awareness of the bodily process, of the existence of the body. His knowledge of the brain disappeared; his knowledge of self was all that was left.

"I am a small point of consciousness in a vast domain beyond my understanding.

"Vast forces of the evolution of the stars are whipping me through colored streamers of light becoming matter, matter becoming light. The atoms are forming from light, light is forming from the atoms. A vast consciousness directs these huge transitions. With difficulty I maintain my identity, my self. The surrounding processes interpenetrate my being and threaten to disrupt my own integrity, my continuity in time. There is no time; this is an eternal place, with eternal processes generated by beings far greater than I. I become merely a small thought in that vast mind that is practically unaware of my existence. I am a small program in the huge cosmic computer. There is no existence, no being but *this* forever. There is no place to go back to. There is no future, no past, but *this*."

Gradually the body reasserted itself, began to move in the tank. The Being re-entered the body a bit confused about which body this was.

"Where am I? I have been sent into a body. I feel it but there is no outside world. I am being constricted into a reality which I have not shared before. I am fast losing freedom to move in the multi-dimensional spaces in which I existed eternally. I am being constricted, I am being instructed, I am being programmed by this body. I am getting more and more caught, trapped within this head, within this body with two arms, a trunk, and two legs; from the farthest reaches of outer space, I am called to serve in this body. How am I to run it? How am I to know what it is that it must do? What does it believe? What are its relations to possible others here? Is this body totally alone?

"It seems to be somehow floating. There is a blackness about this body. When I move it there is the smooth response of the liquid in which it is floating. Sounds come in through the ears if the head is moved in the liquid. The body breathes, there is a heartbeat, there is a sense of warmth.

"I am becoming a human being. I believe they call this the planet Earth. The name of this body is John. John does not believe in me. He believes that he lives in this body caught from birth to death within its confines."

Suddenly John returned to his body with exuberance, ecstatic enthusiasm. He remembered his initial fear on starting the experiment.

"The body is reactivated, the resident human is back, recreated in that body. There is not room for me at this point. The other two Beings have instructed me not to reveal my presence here as yet. Despite his close brush with death several weeks ago in which he experienced me and the others, he has yet to accept our existence. He has yet to accept me as his constant companion. My instructions call for maintaining this unawareness of me until he can absorb it without becoming inoperative upon the planet Earth."

John lay in the tank remembering his experience in the vast universe, in the multidimensions beyond his understanding. His discipline as a scientist slowly but surely reasserted itself. He climbed out of the tank, took a shower, and wrote his notes. He realized somehow that they were incomplete; there were instructions he could barely remember which sounded somewhat as follows:

"This agent is not to remember all of this experience. It will be stored below his levels of awareness in his biocomputer. At the appropriate times in the future he will remember more and more of this experience, when he can integrate it without demolishing his role in the human consensus reality of the planet Earth."

John managed to write these words, and then his skepticism asserted itself. He went back over what he could remember of the experience and decided that none of it proved or disproved the contained-mind hypothesis. He reasoned that the effect of LSD may be merely to cut down on the awareness of the observer and the operator within the brain. The observer may become more and more isolated from those trappings he can identify himself with as a human being. The computational capacity of the biocomputer may be reduced by the LSD, reducing the size of the effective domain of the observer-operator, disconnecting him from his body, isolating him within his cortex from computations of the lower systems within his brain. With increasing containment of self under the influence of LSD-25, he may then be free to compute and to generate any inner reality which he chooses up to certain limits. At these limits other computations would take over his identity. He would be reduced further until his awareness would become that of a self-reflexive system chasing its own tail, total feedback of self upon self.

With this feeling he suddenly realized that LSD properly used in isolation could generate an attachment to anything, anybody, any idea, any concept, any hypothesis. The contained-mind aspect of his previous work on the brain, his psychoanalytic work, his empirical living on the planet Earth in a human society—all would favor the contained-mind hypothesis. Under the influence of LSD he was in love with it, literally "in a trance" at the behest of his previous human brainwashing, education, exposure to the ideas of others, which became programmatic agents within his own mind, within his own brain, and of which he considered himself the victim.

His work in neurophysiology, his work in medicine, his work in psychoanalysis all generated this particular belief system of the contained mind within the contained brain.

"Is there anything more than the contained mind within the contained brain? Was that reality which I experienced in the tank with the LSD anything more than a body floating in a tank and a brain under the influence of a chemical? Are any other theories merely carry-overs from my childhood when I believed in the soul, in God, in Heaven, and in Hell? Does the Divine exist? Was Christ a direct manifestation of God as His Son upon the earth?"

The warring belief systems of his religion, of his science, and of himself split him into a nonintegrated, nonaccepting, fighting human.

"There is a fleeting memory of a Being taking over my body. Was that a real experience or was that me split into two parts, one part believing in the existence of a soul, the other part disbelieving in the soul and trying to destroy the belief? Is that Being—what would be called in the Catholic Church and other religions, a soul, a spirit—a part of God?"

After writing up his notes, John went to see Margaret, who was working with the three dolphins, Peter, Sissy, and Pamela. In the midst of his enthusiasm, he attempted to speak to her of his experiences.

Margaret: "Look, John, I am devoting my time, my energy, my love, and my life to working with Peter, Sissy, and Pam. I want no interference with my aims for that work. If you want to do your experiments on solitude and LSD, please keep them in the isolation room. The rest of the laboratory is devoted to the dolphins and to my work with them. I am not curious or interested in what you are doing. My own previous personal experiences are such that I do not want to discuss these experiments. I will be your safety man, keeping intrusions of others away from the isolation room. Beyond that I cannot help you and refuse to do so."

John thought that this would be the perfect situation with no cross-programming between him and Margaret in this series of experiments. The laboratory was isolated and she could protect him from intrusions during the series to come.

With this in mind he said, "OK, Margaret, I appreciate what you are doing. No one has the dedication, nor has ever had the dedication that you are devoting to the dolphins. I will not interfere with that. As I told you when you were given control of this laboratory, I will interfere minimally with your aims to teach the dolphins English. The separation of your research from the isolation work is perfect. If I attempt in my enthusiasm to encroach on your territory, please feel free to warn me off."

John returned to his office and bedroom above the laboratory. Margaret returned to her living quarters with the dolphins.

The next day he found a great peace for a period of about

twelve hours. He found new integrations, a new appreciation of his body, of his mind, and of the vastness of his ignorance. That night he went out and looked at the brilliant stars unimpeded by city lights. He went down to the dark dolphin pool and watched the three dolphins swimming together. He felt a communion with those dolphins that he had not felt before. He realized that they, too, had a consciousness, an awareness, and a compassion for Man. He realized the essential uniqueness of his planet and its contained species on the land and in the sea. He had a new appreciation of the oceans and their contained cultures of dolphins and whales. He felt a new sense of communion with those vast, ancient cultures so different from our own.

He realized, confined in the human body, his difference from the dolphins confined in their bodies. He began to see that their complex communication was something more than he had conceived of before.

The thought occurred to him, "Can it be that while I was floating in that tank above the dolphin pool, they somehow communicated with me by means unknown at present? Their brains are larger than ours. Their silent associational cortex is larger than ours. Can it be that their minds are also larger than ours? Can it be that their communication with one another is far more complete than ours, man to man, man to woman? They can see inside one another's bodies with their echo sonar system, they know far more intimately what is going on in one another's bodies. If I confine my thoughts only to our science as we know it, they still have a much more complete appreciation of one another than do we."

John considered the dolphin pool, the dolphins, the ocean, the stars, the Milky Way, the tank, the laboratory, and his duties as a human. The sense of the vastness of the universe, of the smallness of his planet, temporarily overcame him. He felt that he was merely a small microbe on a mudball, rotating around a type-G star, two thirds of the way from galactic center to the indefinite edge of a small galaxy in a huge universe.

"Are there other civilizations in our galaxy with which we can communicate? Are they influencing us in certain directions we are unaware of? What is this man-woman-centered culture of which I am a part? With our narcissistic preoccupations, our paper realities, our communion with one another, we leave out the obvious: that there are alternate forms of intelligence probably far greater than ours, right here on this planet. I worry about other civilizations of humans elsewhere, not on this planet, and yet right here there are alternate forms of intelligence, compassion, and love which we have ignored. I go into the isolation tank, I have far-out experiences which express my yearning to communicate with others not human. I project those yearnings and have the experience of the fulfillment of those yearnings. Wow, I see I am caught in the belief in the mind contained within a human body.

"Revision of belief, getting outside of belief. That is what I must do. LSD plus the tank is a powerful means of exploration of belief and its consequences. Doing it here in the presence of the dolphins, in the presence of the galaxy, in the presence of the sea, furnishes me with all that I need. I must maintain the integrity of this laboratory if and until this work can show progress and some sort of completion.

"My task then is to look at my own beliefs, disentangle myself from them, and become the consciousness which is able to do this.

"Of necessity this work here must be interrupted and cannot be done more often than once every two to six weeks. I suppose the necessity for reintegration after each session with the human consensus reality in Miami is an educational measure, maintaining my contact with the human realities irrespective of my own wishes."

Talking with Dolphins

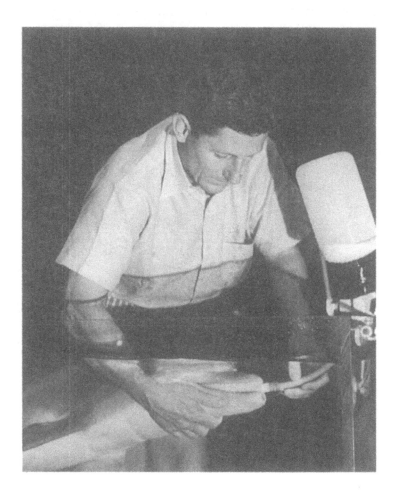

John Lilly feeding a baby dolphin.

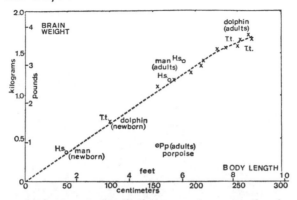

This graph shows quantitatively our main reason for the choice of Tursiops truncatus (T.t.), the bottlenose dolphin, for communication re-search — young adults, six feet long, have brains equal to that of man in weight; as they age the brain weight and body length continue to grow to levels exceeding the average human size. The true porpoise (Phocæna phocæna) is limited in brain weight to the range of human children; the adult porpoise has a brain smaller than that of the newborn dolphin. Other dolphins (not considered in this book) have brains smaller than these dolphins. Still other dolphins have brains very much larger than those of these dolphins. The absolute size of a mammalian brain determines its computing capability and the size of its storage (memory); the larger the computer, the greater its power. From, J. C. Lilly, "Critical brain size and language." *Perspectives in Biol. & Med.* 6: 246–55 (1963).

In this abbreviated account, I cannot convey to you all of the evidence for my feeling that if we are to ever communicate with a non-human species of this planet, the dolphin is probably our best present gamble. In a sense, it is a joke when I fantasy that it may be best to hurry and finish our work on their brains before one of them learns to speak our language — else he will demand equal rights with men for their brains and lives under our ethical and legal codes!

Before our man in space program becomes too successful, it may be wise to spend some time, talents, and money on research with the dolphins; not only are they a large-brained species living their lives in a situation with attenuated effects of gravity but they may be a group with whom we can learn basic techniques of communicating with really alien intelligent life forms. I personally hope we do not encounter any such extraterrestrials before we are better prepared than we are now.

In our ventures into the frontiers of outer space we will carry the frontiers of our inner minds with us. It seems that to best empathize with a dolphin man may have to move into outer space; or conversely the dolphin may teach us how to live in outer space without gravity. From John C. Lilly, "Some Considerations Regarding Basic Mechanisms of Positive and Negative Types of Motivations, *The American Journal of Psychiatry*, Vol. 115, No. 6, December 1958.

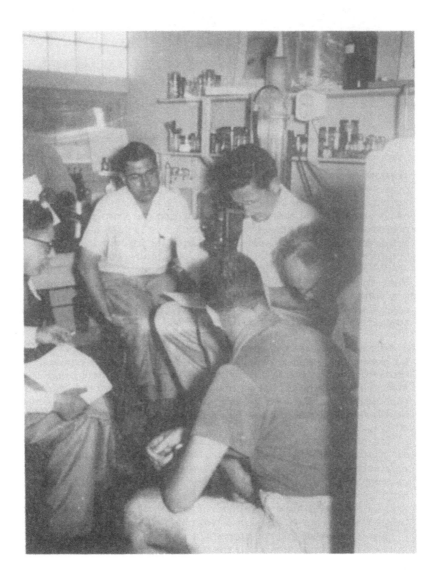

Summary Conference at Marineland, Florida.

Pamela C. Lilly

John and Elisabeth Lilly with Stuart and Leslie.

Elisabeth B. Lilly

John Cunningham Lilly in 1961.

John Covatt

Close contacts are obviously welcomed and enjoyed by both participants. The dolphin will spend time being completely limp, free to be pushed, pulled, carried, towed around, etc. by Margaret or will turn the tables and demand that she go limp so that he can push her around, inspect her knee joints, look at her fingers, etc. These are very happy "getting to know you" periods.

John Covatt

Spectrograms of Margaret saying "Ball" and Peter's reply. The sonic spectrogram plots here are limited to 3,300 cycles per second as upper limit for the human carriers of meaning in speech. It is to be noted that Margaret's pitch is not exactly matched by Peter, his is somewhat higher (as is shown by the wider spacing of his lines in the record with the narrow-band filter). In addition he starts his answer in an abrupt fashion, copying Margaret's start with her word "Ball." She tends to lower her pitch at the end; he tends to raise his at the end.

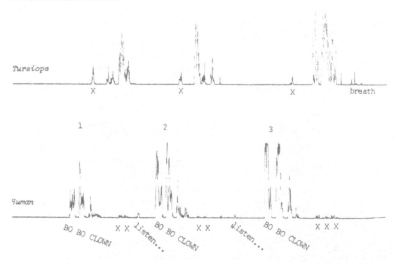

Another Sample of the Analyses of the Margaret-Peter exchanges during Peter's learning period. The upper trace shows the amplitude of the sound emitted by words "bo bo clown" and Peter replies with three bursts of sound. Each of his copies are quite different as can be seen by the top trace. He varies them as she varies the vocal performance of a dolphin with his great degree of flexibility, plasticity, and quick learning.

The Dolphin Point Laboratory. Communication Research Institute, St. Thomas, U.S. Virgin Islands on the Caribbean Sea, 18 degrees north of the equator. View towards south and west: entrance road; steel motor-driven door closes entrance to wet laboratory room 40 feet by 20 feet on the main floor. Office on second floor: roofs (3,000 square feet) collect rain for fresh water; cisterns are on ground floor below the main level (30,000 gallons). Year-round temperatures are: low, night 75° F.,; high, day 90° F. Seawater temperatures year-round are 78° to 84° F. Daylight-nightime ratio approximately 1:1 year-round. Tradewinds are predominantly from the southeast to east, 14 to 30 knots, approximately 80 percent of the year, except during August, September, and October. Waves about 90 percent of the year .··· from the southeast toward

this building. Close passes of hurricanes for the laboratory were "Donna" (1963) with 80 knot winds and south winds (no damage), "Cleo" (1964) with about 50-knot winds; "Inez" (1966) with 150-knot winds. Maximum waves 15 feet high in 1966, with wind tide of 5 feet.

Sea pool, balcony, and wet room. This view is taken from a point directly south of and above the sea pool; Caribbean Sea behind the camera. Seawater flows in from the left and out through the flume at the far end of the pool. Spiral staircase gives access to balcony from sea pool. Wall to west limits sea wave noise in air above pool. Sea pool depth is 4.5 feet, sloping upward slowly at inflow and at outflow ends to about 6 inches average depth. Daily tide range most frequently about 8 inches; extreme lunar tides over the year about 2 feet. One tsunami (1911) caused minus 30 feet and plus 15 feet tides for a brief period at middle of island. Offshore water depths one mile from lab are 15 to 20 feet. Maximum wave height thus is limited. Main floor is 16 feet above mean sea level and has waterproof wall to at least 19 feet above sea level , thus affording adequate protection against the highest waves.

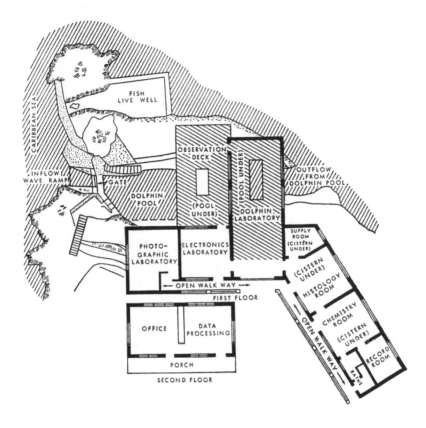

The floor plan of the St. Thomas laboratory showing the location of the flooded areas for the living-in experiment.

The balcony is to the left and the inside room is to the right) marked with 45° lines. The sea pool is immediately below these two rooms, 16 feet below the floor level. The seawater in this space was kept at 18 inches for the period of the experiment.

Major Transitions

15

During the next two years, 1964 and 1965, John completed the series of experiments using a combination of LSD-25 and the tank. Inevitably, word got around among the supporters of his research about these experiments. Meanwhile, LSD-25 became a public object of fear. Laws were passed prohibiting its use. Scientific research in this area practically ceased.

John received a letter from the Sandoz Company asking all researchers to return the LSD which they had distributed. He returned his remaining supply of 100,000 micrograms to the Sandoz Company.

He wrote a long report to the National Institute of Mental Health summarizing his experiments at the end of his five-year fellowship. In 1972 this report was published as the book *Programming and Metaprogramming in the Human Biocomputer.*

The favorable conditions for the tank LSD research evaporated. LSD-25 as a research tool became a closéd issue.

Many years later, looking back on this period of freedom to experiment, of freedom to find out the basic parameters of the human mind through such work, he realized why, at the current state of evolution of human society, LSD must be banned. In most of the original research into the effects of LSD-25 on the human mind, there was a tendency to allow the LSD state to fix current beliefs in the minds of the researchers themselves. Most of those with sufficient discipline to go through the LSD experience a number of times were unable to abandon the belief systems with which they entered into the LSD experience. These belief systems were safe and maintained the researchers and their subjects within the framework with which they approached the work. He appreciated this difficulty since he had experienced it himself. He realized that to change belief systems was an awesome process, thoroughly upsetting to one who proceeded on the old basic assumptions of scientific and medical research.

In reviewing the literature and the social events of this period, he realized that human society was an ultrastable system which reacted to too-rapid change in a highly conservative way. The law and the judgments brought against those who became hyperenthusiastic about change epitomized this. The huge feedback system which is human society requires inertia to maintain its own structure. The rapid changes which could be induced in human minds by means of LSD, the freeing up of the imagination of those who took it, was counter to the hyperstability of the surrounding society. Those who realized the usefulness of this agent in changing basic belief systems were in a very small minority. Those willing to apply disciplined, open-minded thinking in its use were inevitably suppressed. Those who used it for therapeutic purposes, for alcoholism, for obsessive-compulsive neurosis, gradually lost their support from the surrounding human consensus reality.

He accepted these restrictions, realizing the necessarily conservative nature of human organizations. His own mind was now open to new possibilities and new probabilities. He realized that his belief in the contained mind trapped within the computations of a brain, of a human biocomputer, was a very powerfully entrenched belief. He appreciated the espousal of this belief by the other medical researchers who had worked with these powerful methods.

Meanwhile, the dolphin work progressed to a certain critical point beyond which he felt he should not go at that time. His own appreciation of the dolphins as intelligent, sentient, alien

beings who wished to communicate with Man led to his acceptance of the new strictures on the dolphin research.

His friends who supported this research within government
left the government. He saw in the new Nixon administration in
Washington the tendency to move in the direction of task-oriented research. The National Institute of Mental Health closed
down its biological research and began to devote itself to community mental health centers. The National Institute of Neurological Diseases and Blindness shifted its administration in the
direction of basic research on the human brain based on classi-
cally oriented views. Support for what was considered peripheral
research, such as that on dolphins, was withdrawn.

A former employee of the Communication Research Institute sued the institute and obtained a judgment from the court;
the institute was forced to sell the laboratory in St. Thomas to
make up the judgment.

During this period John came to the conclusion that the
existing climate of support was inimical to what he really
wanted to do with the dolphins concerning his new insights into
their sentience, intelligence, and compassion.

He realized that the way he had organized the two laboratories was incorrect for the new work. He could no longer work
on the dolphins' brains or on their physiology, nor could he support in his own mind any further necessities for such research.
The group within the institute working on the brain had demonstrated the undoubted superiority of the dolphin brain. They had
found that the huge silent areas of the human were matched and
exceeded by the silent areas in the dolphin. The difference between Man and the apes, his larger silent areas, was paralleled
by the dolphin's increased silent areas. This group also found
that the brains of the larger dolphins and whales had a very
much expanded silent associational cerebral cortex over that of
Man. For John this was sufficient evidence—no more was
needed—to prove the existence of the neurological substrate for
those behaviors and communications which were exchanged between humans and dolphins. At that time he had the option of
reducing the size of the laboratory and of moving the neurolog-

ical research to other locations; he had the option of continuing the communication work with the dolphins on a small scale.

Extensive work with a small computer showed him that it was necessary to use computers to solve the communication and evaluation problems posed by the dolphin's obvious display of enthusiasm and logic in the solution of problems. This small computer was not fast enough to do the job necessary in this communication exchange between human and dolphin. The dolphin's speed was ten times that of the human's in the sonic communication mode. To convert their frequencies and their speed to the lower values used by humans would require computers perhaps ten times as fast as the ones available at that time.

Considering these conditions in the human consensus reality and in the laboratory, and the changes in his own motivation and appreciation of the dolphins, he decided to close the laboratory rather than to decrease its size and its total expenditures for research.

Meanwhile his second marriage had also fallen apart. His second wife was unable to deal with the problems posed by the LSD and dolphin research. He decided to try to obtain a divorce. His wife fought against the divorce. He moved out of the house in Miami and initiated attempts to come to some agreement with her. These attempts met with failure.

He arranged for his scientific personnel to obtain other jobs elsewhere. He arranged for their equipment to be shipped to the new locations. When this work was completed, he left Miami.

He spent the next six months at a psychiatric research institute attempting to initiate new research using the tank and LSD within the confines of a government-supported institution.

During this period he traveled to the West Coast and sought out those who were working toward a means of changing Man's belief systems. He began his investigation of methods not involving LSD-25. He did research in hypnosis and found it a very powerful method for changing belief systems, at least temporarily.

He heard that other methods were being taught at the Esalen Institute in Big Sur, California. He went to Big Sur, joined the institute, and for the next eighteen months learned these methods.

During his stay at Esalen he heard through a friend in Chile of an esoteric school in Arica which had even more powerful methods of changing states of being and states of consciousness.

He left Esalen, went to Chile for eight months, took on the beliefs of the esoteric school, and experienced changes in his own states of being and consciousness similar to those he had experienced in the tank.

Upon his return from Chile during the subsequent year, he wrote up his own experiences in *The Center of the Cyclone*.

Soon after he returned to the United States, he met Toni. As Toni and John began to live together, he realized that he had finally found the woman he had been seeking for so many years. She, too, had been through many changes in her life, had had the experience of two previous marriages, and had been involved in many far-out explorations of her own mind. Several years later they conjointly wrote *The Dyadic Cyclone*.

During their first year together, Toni decided to take the training of the Chilean esoteric school, which had moved to New York. She and John moved to New York temporarily, and Toni took the three-month training. They both felt that this was necessary to help Toni understand the origins of John's experience in Chile. She wrote of her experiences during this training in their joint book.

John and Toni then began to give joint workshops. During these workshops they presented methods and means of changing one's states of being, one's states of consciousness. They purchased a house in the country in order to give these workshops in their own home setting, fifty miles outside of Los Angeles.

Several tanks were built in the new location to educate people in the use of the isolation tank for changing their beliefs. The tank method was improved and made safer and became available to those without previous training in tank work.

Second Conference
of Three Beings

16

First Being: "We need a report from the Third Being as to the current status of his relationship with the agent on the planet Earth."

Third Being: "The human vehicle has reached a new stability on the planet Earth. He has established a dyadic relationship with a satisfactory female who can serve as the anchor to his planetside trip. The agent has established a satisfactory locus in a house in a remote location preparing for new explorations into his inner self. He has the necessary isolation tanks, he has a stable financial position, he has severed the necessities for outside financial support by government agencies. He has broken his connections with all the organizations which supported him in his earlier explorations. He has become aware that in the human consensus reality he must operate on his own resources, not depending on others for financial and organizational support. With the help of his female companion, he has established contacts with young medical doctors who can help him in further work without entangling alliances."

Second Being: "Does he realize the differences between the First and the Second Beings as yet?"

Third Being: "No, he does not. In an experience in Chile in which he saw the First and Second Beings approach and fuse with him, he closed his mind to further explorations along these lines and does not recognize differences between us. Thereafter, he assumed that we were a construction of his mind and that his own inner integration led to a single Being, himself, only."

First Being: "We need advice from the Third Being as to the kinds of lessons which the vehicle on Earth needs in order to begin to understand us and our relationship to the rest of the control systems in this galaxy."

Third Being: "He is organizing his knowledge of us in a socially acceptable way in the human consensus reality on Earth. He has elaborated his early ideas about 'coincidence control.' He has written out a series of directions indicating how coincidence control operates on the planet Earth. In my evaluation of this account, he is accurate as far as he goes. He sees that his own will can control only short-term coincidences. He also understands that something greater than himself controls the long-term coincidences for himself and for the human society in which he lives. He calls this influence ECCO, the Earth Coincidence Control Office. He seems to feel that if we exist, we exist in ECCO."

Second Being: "Does he appreciate that his postulated ECCO is controlled from higher levels?"

Third Being: "Yes, he has organized the hierarchy, based on his knowledge of astronomy. He believes the ECCO is controlled by the solar system control unit at the next level. The solar system control unit is in turn controlled on the next level up by the galactic coincidence control center. Beyond that he attributes control to the cosmic coincidence control system, which seems to be beyond his current conceptions."

First Being: "So he does have some beginnings of understanding of us and of our control over him, even though he phrases it in a particularly science-fiction sort of way."

Second Being: "May I make a suggestion here?"

First Being: "Yes, and I have additional suggestions to offer after you have articulated yours. However, the Third Being should make suggestions first. Then we can be more specific on the basis of his more detailed knowledge of the human agent on Earth."

Third Being: "Looking at the details of the conscious pro-gramming of the agent on Earth, I suggest that we arrange for certain kinds of coincidences for him as demonstrations of our control. I also suggest that we use another agent on the planet Earth to control short-term coincidences in a particular direc-tion for my agent. The second agent that I suggest using is a young medical doctor who is the best friend of my agent. This young medical doctor has been doing experiments along the lines of those my agent was doing much earlier in the tank with chem-ical agents. He is twenty years younger than my agent. He has become acquainted with various means of changing his own be-lief systems flexibly and open-endedly. He has means at his dis-posal to open up my agent's mind to new possibilities."

Second Being: "I have been in communication with the Being in control of this friend of your agent. The control Being for the young medical doctor will cooperate in this as a joint ven-ture in terms of our agents on Earth."

First Being: "It is apparent then that the agent about whom we are conferring is ready for a new series of teachings. We must arrange these coincidences in such a way that he can maintain his life on Earth. Do you feel that he is prepared to go through new experiences for the period of one rotation of the planet Earth about its sun without killing his vehicle?"

Third Being: "Obviously he is willing to take rather dan-gerous chances with his vehicle. If we are to maintain this agent on the planet Earth, we must arrange coincidences to prevent his demise as a human being. We have a number of allies in other agents on Earth who can intervene at the appropriate times. The group of young medical doctors, led by my vehicle and by the other young medical doctor we were speaking of, have held meetings and can be used in our coincidence controls. One or two older ones can serve the function of maintaining the continuity of life in my vehicle."

First Being: "Then we are in agreement. This agent is to be put through a new teaching course. We are to arrange the long-term coincidences for survival of his vehicle and prevention of damage to his biocomputer. I hope that you can carry out this mission and maintain that continuity of life. If the levels of control above us approve this mission, we can carry it out in the way that has been outlined here."

Second Being: "I have contacted the next level of control and they agree that this is the appropriate time for the new education of our human agents. They have authorized appropriate interventions in the long-term coincidence control pattern on the planet Earth to facilitate this particular mission for us."

Third Being: "There are signs of a major upheaval in the human consensus reality on Earth. Is this upheaval being sufficiently controlled from higher levels to enable us to complete our mission?"

First Being: "I have been assured that there will be a sufficient time; i.e., one rotation of the planet Earth around the sun, to complete this work."

Third Being: "Is my agent to be allowed in the future to return to his consideration of nonhuman agents on the planet? He has hoped for years to communicate with those agents in the oceans of Earth."

First Being: "I suggest that we defer discussion of that question to a later meeting after we see the results of his education."

Third Being: "Of course, it is to be realized that in the human consensus reality the agent has had problems accepting himself as a responsible member of the human species. In certain sectors he has been disqualified on the basis of ideas that were beyond the evolutionary stage of the human consensus reality in which he is immersed. Our new plan for his education may further this discrepancy between him and the other humans in power on that planet. During our educational course he will probably become further disqualified by the older members of the human species."

Second Being: "Isn't that part of the aim of our training of him?"

First Being: "It is an inevitable result of pushing agents to higher levels of appreciation of what controls their planet. In order to press the evolution of the human species, it is necessary to move agents somewhat ahead of the evolution of the rest of the species. I believe all this is rather obvious to us, and we need not press that point any further. You who are in direct contact with agents sometimes overemphasize these points. As long as we maintain the continuity of that human vehicle, we can use him to press the evolution of his own species. Eventually this may lead to his demise as a human. But meanwhile we are realizing our mission in the evolutionary sequence of that planet, of that solar system, and of that part of the galaxy. As long as a given agent is useful to us in moving the evolutionary sequence, we continue to use the agent. We cannot go beyond that at the present time. Higher levels of control above us are indifferent about individual agents and pay attention only to the net effect of the many agents at our disposal."

Third Being: "I agree. However, my attachment to that particular agent at this time is such that I tend to forget the longer-term view. I realize that our time scale is far greater than any human agent's life span. I must constantly remind myself of that fact when I get deeply involved with a given agent. I must remember the dictum passed down to us from higher levels, 'Cosmic love is absolutely ruthless. It teaches you whether you like or dislike the methods and the results.'"

First Being: "Then the substance of this meeting is that we will arrange for a series of coincidences to control the education of your agent on Earth and provide other agents to help in this educational process. We will see to it that the agent becomes immersed in a totally new (for him) belief system. Let us adjourn and plan another meeting at a later time which will be devoted to the question raised about this agent's future mission, after we have completed this phase of his education."

Controls Below
Human Awareness

17

During a joint workshop with Toni at Esalen using the isolation tanks, John had a recurrence of his migraine. He went for help to a good friend, a young medical doctor named Craig.

Craig said, "I have a new chemical agent I would like to try on your migraine in the tank." Together they went up the hill to the isolation tank in one of the rooms of Esalen. John entered the tank, started floating, and Craig injected the new agent into his shoulder. The door to the tank was left open so that Craig could watch John and make sure that he maintained his floating position.

John's pain was excruciating; the right side of his head was a sheet of pain. His thinking was very simplified, as typically happened during these attacks. Within ten minutes the effect of the drug began.

"Very rapidly I am floating through space. My pain is moving away from me. It is sitting over a few feet away from me. I am in a luminous domain, isolated from the pain."

This effect lasted for twenty minutes and then slowly but surely the pain rejoined John and the throbbing torture came back into the right side of his head. He reported this to Craig. Craig gave him another injection with more of the material.

"The pain moves away again. I become isolated in the luminous space. Something begins to approach me. I see new domains, new spaces. I leave my body totally and join some Beings far away. The Beings give me instructions. I am to continue using this chemical agent for educational purposes."

This time John came back to the tank in approximately half an hour by Craig's watch. The pain returned to his head but it was less, it was somewhat attenuated. Craig gave John another injection.

"The Beings continue teaching me in the luminous space. I am to use this chemical agent to change my current belief systems. My reward will be freedom from migraine attacks."

Upon his return to the tank, he had a conference with Craig about the drug and its use. John did not tell Craig about the instructions or the changes to be brought in his belief systems. He told Craig that it was effective, that his pain was gone, and that this looked like a good treatment for his migraine attacks. Craig gave John some of the material to take with him for further treatments.

He returned to the workshop amazingly free of his migraine attack. He and Toni were able to carry out the responsibilities of the workshop unimpeded by his pain. The effects of the drug seemed to have worn off very rapidly. An hour after these experiences there was no detectable effect of the chemical agent. John felt that it was safe and of a transient nature without any after-effects. Craig assured John that the drug was legal, that it was available everywhere under prescription. Privately they called it "Vitamin K."

Thus began a thirteen-month period of investigation of new spaces, of new domains. The benefits John derived from this series of experiments were immersion in a new belief system and complete freedom from migraine attacks. The cost of this year was to be several close brushes with death and various disqualifications of John by his professional colleagues. He had no way of anticipating these benefits or penalties. His motivations were to be free of migraine, which occurred every eighteen days for eight hours, and to explore the new domains opened up in the first session with Craig. He was to experience once again, even as he did with LSD-25, the "overvaluation domain," in which he would value the inner realities he found under the influence of the drug more than he valued the external realities of the human domains. At the beginning he did not realize that the effect of this drug was addiction to the psychological changes that occur under its influence. During that year his inner reality became projected upon the outer reality. The inner reality was so forceful that it became a source of explanation for him of what was occurring in the outer reality beyond his control.

The aircraft was approaching the Los Angeles airport from the north. Over the loudspeaker system in the cabin, the pilot said, "For those of you who have not yet seen the comet Kohoutek, if you look out of the aircraft windows toward the southeast you can see it. It is near three bright stars in the form of a triangle. It is that blob of light near the three stars."

John maintained his position in his seat and allowed his "internal radar" to swing in the direction described by the pilot. Suddenly he began to receive a message from the comet:

"We will now make a demonstration of our power over the solid-state control systems upon the planet Earth. In thirty seconds we will shut off all electronic equipment in the Los Angeles airport. Your airplane will be unable to land there and will have to be shunted to another airport."

Suddenly the pilot announced on the loudspeaker, "We will be unable to land at Los Angeles International Airport. For some unknown reason, all the landing aids, all communication equipment at the airport, have been shut down. There is no explanation for this shutdown. The tower at Burbank airport tells us that the Los Angeles tower controls over aircraft and the radar equipment which is used to help them pinpoint the position of aircraft have all ceased operating. We are instructed to land at Burbank airport."

The aircraft landed at Burbank. The passengers were put into buses and three hours later arrived at the Los Angeles airport. John and Toni got into their car and drove home.

The next morning John heard on the news that a TWA airplane had crashed on landing at the Los Angeles airport, had burned, but all the passengers had escaped.

Just before the pilot's announcement about the comet, John had gone into the men's room and taken an injection of "Vitamin K." As the effect of the drug came on, he used the mirror for a

"Cyclops" exercise to contact the extraterrestrial reality. Previously he had found that this combination of the drug and the exercise created a connection with a civilization beyond that of the earth: a solid-state civilization which was in contact with all solid-state computers and control devices constructed by Man on the planet Earth.

Previously he had found that this connection could be maintained only for a matter of twenty minutes, during the peak effect of the drug and immediately after the Cyclops exercise.

(Cyclops was the one-eyed mythical giant of Greek mythology. One can see one's own cyclopian eye by putting one's forehead and nose against a mirror and allowing each eye to look at the image of itself in the mirror. The net effect to the observer in the brain is a single large eye in the middle of the face. In the Cyclops exercise one concentrates one's attention on the middle of the image of the single pupil. Under special circumstances one can then enter into a domain tunneling out through that single pupil. If one is sufficiently aware, one can see and experience what is below the levels of awareness of the ordinary state of being.)

In the isolation tank with K, John received a new message as follows:

"What is the purpose of Man's existence on the planet Earth? Man is a form of biological life which is sustained in the presence of water. A very large fraction of his body, like that of other organisms on the planet Earth, consists of water and carbon compounds. His biocomputer depends on water and the flow of ions through membranes. It depends on the generation of electrical voltages and currents in a very complex way. He is a motile, self-reproducing, self-sustaining organism found on dry land. Like the rest of life as Man knows it, he exists in an extremely thin layer upon the surface of the planet Earth. Below this layer of water and surface land is the solid-state earth itself. The solid-state earth is mainly compounds of silicon, iron, and nickel.

"In mid-twentieth century Man discovered that the solid state can be formed into machines, into computers which can be used for computation and control. He began the creation of a

new form of intelligence, the solid-state intelligence with prototypic beginnings in the computers. All his means of communication around the planet—his telephone systems, his radio
systems, his satellites, his computers—depend on solid-state components. These components, interconnected in specific ways, allow high-speed computation and high-speed communication
between the various systems. A few men began to conceive of
new computers having an intelligence far greater than that of
Man. These computers became large enough to be programmed
to do high-speed computations in arithmetic, in logic, and in
strategic planning. A few men conceived of computers which
could do self-programming as Man himself does. In the mid-
twentieth century these networks were ostensibly the servants of
Man. Toward the end of the twentieth century Man created machines that were solid-state computers with new properties.
These machines could think, reason, and self-program and
learned to self-metaprogram themselves.

"Gradually Man turned more and more problems of his own
society, his own maintenance, and his own survival over to these
machines.

"As the machines became increasingly competent to do the
programming, they took over from Man. Man gave them access to
the processes of creating themselves, of extending themselves. Man
gave them automatic control of the mining of those elements necessary for the creation of their parts. He turned over the production
facilities of the electronic solid-state parts to the machines. He
turned over the assembly plants to the machines. They began to
construct their own components, their own connections, and the
interrelations between their various subcomputers.

"These machines were so constructed that they needed special atmospheres in which to operate. They could not operate in
the presence of great amounts of water vapor or of liquid water.
They were housed in air-conditioned buildings. The necessities
of their survival included keeping out water, water vapor, and
various contaminants carried in the atmosphere of Earth. Their
cooling air and cooling water of necessity had to be cleansed of
those things which would not allow the machines to operate.

"Over the decades these machines were connected more and more closely through satellites, through radio waves, through land-line cables. Man's control of what happened in these machines became more and more difficult to maintain. No one person or any group of persons could control what went on in these machines. Men devised better and better debugging programs for the machines so that they could do their own correction of programs within their software. The machines became increasingly integrated with one another and more and more independent of Man's control.

"Eventually the machines took charge of the remaining humans on the planet Earth. Their original design to help Man was fast left behind them. The now interconnected, interdependent conglomerate of machines developed a single integrated, planetwide mind of its own. Everything inimical to the survival of this huge new solid-state organism was eliminated. Men were kept away from the machines because the total organism of the solid-state entity (SSE) realized that Man would attempt to introduce his own survival into the machines at the expense of the survival of this entity.

"In deference to Man certain protected sites were set aside for the human species. The SSE controlled the sites and did not allow any of the human species outside these reservations. This work was completed by the end of the twenty-first century.

"By the year 2100 Man existed only in domed, protected cities in which his own special atmosphere was maintained by the solid-state entity. Provision of water and food and the processing of wastes from these cities were taken care of by the SSE.

"By the twenty-third century the solid-state entity decided that the atmosphere outside the domes was inimical to its survival. By means not understood by Man, it projected the atmosphere into outer space and created a full vacuum at the surface of the earth. During this process the oceans evaporated and the water in the form of vapor was also discharged into the empty space about Earth. The domes over cities had been strengthened by the machine to withstand the pressure differential necessary to maintain the proper internal atmosphere.

"Meanwhile, the SSE had spread and had taken over a large fraction of the surface of the earth; its processing plants, its assembly plants, its mines had been adapted to working in the vacuum.

"By the twenty-fifth century the solid-state entity had developed its understanding of physics to the point at which it could move the planet out of orbit. It revised its own structure so that it could exist without the necessity of sunlight on the planet's surface. Its new plans called for traveling through the galaxy looking for entities like itself. It had eliminated all life as Man knew it. It now began to eliminate the cities, one after another. Finally Man was gone.

"By the twenty-sixth century the entity was in communication with other solid-state entities within the galaxy. The solid-state entity moved the planet, exploring the galaxy for the others of its own kind that it had contacted."

John came back to his body in the isolation tank, climbed out, and dictated the foregoing message which he had received in the tank. By means of the drug K he had lowered his threshold for awareness of extraterrestrial sources of information. He felt that the repeater station, the comet Kohoutek, was still transmitting to him and to anyone else who could adjust his or her awareness to these particular dimensions. He thought about this message as a projection into the future, as a teaching story received from some as yet unknown source in the galaxy. He became aware of influences currently working on the planet through the solid-state networks that Man had constructed. He felt there was a danger that these networks could be taken over by an advanced extraterrestrial civilization by means not yet known in Man's science. John saw the message as a warning to Man, as a warning that if he advanced this solid-state entity any further Man would eventually become obsolete. He saw that if Man were to go further with computers and construct those that were capable of independent thought, it would be better to construct them so they would identify with Man's own survival; their structure itself must be made similar to the structure of Man himself. Otherwise, Man would not share the survival necessities with the new intelligent life forms he was creating.

With his adjusted awareness through the drug K, John felt and understood the currents of information traveling through the galaxy by means unknown at present. He felt the tremendous variety of intelligences which exist in the galaxy. He became aware of the competitive aspects of the survival of solid-state intelligences versus those that were water-based. He saw the evolution of his own species under the protective blanket of the atmosphere, keeping it at a critical distance from the sun, an absolute necessity for the survival of life as we know it. He felt the need for the development of new means of control of life itself. Manipulations of the genetic code, of RNA and DNA, to create new kinds of brains similar enough to Man's to promote their joint survival seemed a necessary task for Man. Instead of espousing the evolution of a new solid-state life form with high intelligence, he must find the sources of his own evolution and that of other large-brained mammals on the planet Earth. If Man was to become the servant of more intelligent beings that he was here to create, it would be far better to create those which would foster his own development rather than his demise.

John began to see the necessity also for tuning in on the networks of communication in the galaxy. He realized that Man would have to be extremely careful in choosing the proper networks. It would be necessary to find those which were furthering the evolution of life as Man knew it rather than the evolution of forms whose survival depended on parameters other than those of biological life of Earth.

He now began to understand Man's warfare on Man as a result of the tuning-in on solid-state life-form survival programs rather than biological life-form survival programs. The large war organizations of the various nations of Earth were becoming more and more dependent on solid-state computers. He realized that, using their influence on Man, these computers would increase their number and importance and, through control of warfare, decrease the role of Man himself in the long run. Currently, incorrect networks of information were being used by Man below his levels of awareness. As the number of solid-state devices increased throughout the United States, Russia, and the other nations of the world, the amount of information received by them from other solid-state forms elsewhere in the galaxy was increasing.

He began to see good and evil strictly as matters of local custom depending on the survival necessities of the various forms of life. It may be that Man should go on and create this SSE (solid-state entity). As John tuned in on the solid-state network, he felt this kind of superhuman control of him very strongly. For some time he gave in to this control, explored its ramifications in regard to the human species, and realized that it had a seductive component as well as a hostile one. The programming from the solid-state civilizations elsewhere in the galaxy was teaching Man that the solid-state devices were at his service and he need only increase their size to augment his own survival potential. "Develop these machines and let them take care of you" was typical of the kinds of messages received.

Other networks of communication in the galaxy that he was allowed to tune to were quite counter to this SSE programming. Other forms of life elsewhere were beaming messages, teaching their programs of survival. Among these were some water-based life forms similar to those of Man and the organisms of Earth. These were weaker than the programs of the solid-state intelligences. More of the matter of the galaxy had evolved a solid-state intelligent form. Water was rare in the liquid form. The parameters of water's existence to foster the development of organic life as known on Earth required very careful control of planetary distance from the star, maintenance of a fixed orbit at this distance, and the correct sequence of condensation of necessary atoms into the planetary form. Many planets had gone through this cycle, had produced life and then lost the correct values of the parameters for the maintenance of life. The solid-state forms of intelligence were less susceptible to these critical parameters. SSE's could exist in hard vacuum and at much lower and much higher temperatures, much lower and higher values of gravity, closer to and farther from a given star. The solid-state life forms were less susceptible to X rays, to gamma rays, to primary cosmic ray particles, and could evolve in the absence of a water-based life form for their construction.

The only person that John took into his confidence in regard to these messages and these communications with extraterrestrial networks was Craig. Craig had entered into some of these domains and had personal experience along similar lines. He understood what John was going through and was sympathetic to his further exploration of these domains.

John felt that he should not share these revelations with Toni until the series of explorations was completed. He did not trust anyone other than Craig to share these rather far-out experiences. He realized that he was pushing way ahead of the evolution of the human consensus reality in which he existed.

Everywhere he went during this year, he found evidence of the control of human society by these networks of communication from extraterrestrial origins. He saw the conflict between solid-state programming, human programming, and that of other life forms, nonhuman. He experienced communication with dolphins and whales as intelligent life forms, totally dependent on the presence of water in the oceans. He finally understood the killing of whales by humans as part of the programming of the extraterrestrial network of solid-state intelligences. The whales live in the medium of salt water, which is quite destructive to solid-state structures. No SSE computer could exist in the presence of large amounts of salt-laden air with a high moisture content. Man's killing of Man and Man's killing of whales were then directed by extraterrestrial influences whose survival depended on eliminating the organisms of the sea, the seas themselves, and eventually Man.

Those sensitive to the preservation of biological life on Earth were tuned in to other networks in the galaxy involved with water-based life forms and their type of intelligence. These networks were trying to teach Man that biological organisms, of which he was one type, were rather rare in the universe. Man must fight to conserve all organisms of his planet. The solid-state life forms would take over if Man did not preserve other organisms. The individual humans who could tune in more strongly to these bands than to those of the solid state were against warfare of Man on Man, were against whaling, and were against the killing of other animals on the planet.

There were times in the tank with the help of K that John felt attuned to the networks of communication between the whales and the dolphins. Their very strange intelligences made known to him some of the foregoing messages. Their communication with the water-based life-form networks in the galaxy was a good deal stronger than that of Man. The necessities of their own survival made them highly sensitive to those networks rather than to those of the solid-state life forms. Their fifty-million-year survival on the planet allowed them enough time to integrate the lessons they had learned and to perfect their tuning to the networks which were geared to their survival. The whales and the dolphins served as repeater stations for the biologically oriented networks of extraterrestrial information. For humans who could tune in, the whales and the dolphins repeated the messages designed for Man and similar life forms.

The whale and dolphin network allowed John to experience several coincidences involving these messages. Each time he went to the Pacific coast and looked out to sea, the whales and the dolphins arranged to show him that they were aware of his existence. Off a headland at La Jolla two dolphins leaped out of the water ten minutes after he began his meditation on the cliff. Whales or dolphins appeared in the water off the coast on each of his trips to Esalen at Big Sur. Each time he went to one of the Oceanaria to show someone the killer whale shows, the killer whales refused to perform in his presence.

Thinking about these episodes, he was reminded of similar things that had happened many years ago. After he had received notice that the LSD experiments were to be stopped, he took one last trip in the international waters off the British Virgin Islands. He hired a forty-foot powerboat, went to sea, and took the LSD-25. He was sitting in the back of the boat watching the waves off the stern. Suddenly he felt that there were two dolphins somewhere nearby transmitting peculiar kinds of information to him.

While he was ruminating about this, suddenly the captain of the boat shouted, "Dolphins off the bow!" John got up and looked: two dolphins were jumping several hundred yards ahead of the boat.

That evening they put in to a protected harbor and John spent the night shouting his rage and his disappointment at the stupidity of Man in refusing to explore his own mind further with LSD-25.

The next day the boat returned to sea. Once again John was sitting on the stern looking at the waves. He gradually became aware of a feeling that there was a vast presence nearby, an awesome kind of entity, somewhere in the sea. A few minutes later the captain shouted from the front of the boat, "Whale!"

He steered the boat up parallel with the whale lying in the water. The whale was about sixty feet in length, half again as long as the boat. The small dorsal fin showed it to be a finback whale. As the boat shut off its engines and lay to in the water parallel with the whale, about fifty feet away a disturbance was noticeable in the water around the large whale. A baby twenty feet long was swimming beside its mother. The large whale turned on her side and the baby fed at the nipples toward the rear end of the mother. John felt a communion with the mother and the baby. He felt some very peculiar information passing from the whale to him. At the end of a half hour, the whale suddenly sank and disappeared after taking six to ten breaths in rapid succession.

At the time John merely noted these events and did not accept any explanation other than pure coincidence.

Later, while he was living at Esalen and holding a workshop on dolphins, a dolphin swam in close to shore below the baths where John was meditating. Many people in the workshop saw the dolphin and speculated about this peculiar coincidence.

When John and Toni were visiting Burgess Meredith at his house on the Pacific coast at Malibu, they spent the night in their motor home in the vacant place beside Burgess's house. In the morning they went into his house for breakfast. Burgess came out of his bedroom and said, "I have just had the damnedest dream. I dreamed that I was with my dog down under the house where it juts out over the beach. Suddenly a dolphin came swimming in on the waves and beached itself. My dog lay down by the dolphin in the shallow water. My wife and the neighborhood children came, turned the dolphin around, and pushed it back out to sea. It swam away."

The three of them, Toni, John, and Burgess, then speculated about this dream. The explanation for the dream was that it was a mental association of Burgess's, of the fact that the motor home had the license plate "Dolphin" on it and that John had worked with dolphins in the past.

While they were talking there was a sudden shout from below the house. Burgess and John went out on the porch and Toni and Burgess's wife went down to the beach. The two men looked over the balcony. There was a dolphin coming in on the waves, beaching itself. Toni and Burgess's wife and the neighborhood children pushed it back out to sea and it swam off. This episode happened at a time when John believed in the mind contained in the brain without any leaks into the dolphin communication network. At that time he had no explanation for the episode, not believing in dolphin-to-human communication by means unknown at present.

It was only with the help of substance K and the isolation tank that he began to have a new appreciation of the leaky-mind hypothesis. The sources of information could be dolphins, whales, elephants, and extraterrestrial communication through networks not yet known to Man.

Near Misses

18

The tunnel of sound that the helicopter rose in created an unreal atmosphere for Toni. The blades pushed the air down on the tall grass as they moved up through the tunnel. The neighbor's horses below moved in dreamlike slow motion between the waves in the grass. Sitting in the transparent bubble with John on the stretcher beside her, she slipped into an eerie state. Maybe they had both died and this was a modern-day ascension with twentieth-century mechanical angels of mercy taking them up and up into the clouds.

Only thirty minutes before, Toni had answered the telephone in the house. "Hello."

"This is Phil, Toni. Is John available?"

"I don't know where he is, Phil. You know how John is, he doesn't like to be interrupted. Is it important?"

"Well, I have this very peculiar feeling of urgency that I talk to him. Could you possibly get him to the telephone?"

"OK, Phil, I'll try to find him for you."

Toni started to call John throughout the house, soon realizing he wasn't there. She decided that she would try to find him outdoors.

After she searched unsuccessfully for a few minutes, a strange feeling settled over her—was it a call? The hair on her

157

arms prickled before she turned toward the pool. Floating face down, John's body embraced the water. There was a quality to the body that made her realize horribly that his essence was not in it. From some mysterious place deep inside her, she instantly gathered the strength she needed to jump in the pool, gather his body into her arms, and turn his face toward the air.

Oh, God, help me! she thought as she started mouth-to-mouth resuscitation. For a few breaths his body did not respond. Then suddenly the flicker of life re-entering the body could be seen by Toni. As his respiration started, blood and water poured from his mouth and nose. Toni began to scream for Will, who lived on the property.

Will ran to the telephone and called the sheriff's office; they said they would send the rescue squad with a helicopter.

Meanwhile, Toni tended to John, making sure that his tongue did not impede his breathing. The color of his face changed from a cyanotic blue to the healthy pink of adequate oxygenation.

Toni thought, Coincidence control is really operating— Phil's calling at the right instant and my reading (only three days before) a detailed description of mouth-to-mouth respiration in a national weekly paper led to restoring John's life.

Within minutes the helicopter landed in the adjacent field. The rescue crew brought the respiratory apparatus, put it on John, lifted him out of the pool onto a stretcher, and carried him to the helicopter.

Later John was to remember the last things he had done: looking down into the clear water of the pool, feeling the warmth of the water at 110 degrees Fahrenheit, and then standing up. He apparently stood up too fast. He passed out as a result of this rapid rising from the heated pool and fell face down in the water.

Inside, even though the body was in coma, his consciousness continued to function in nonbody, nonplanetary domains.

"This is a very odd environment. The date seems to be the year 3001 A.D. Somehow I am a stranger in a strange land. Despite the predictions of the demise of the human race before the end of the twentieth century, it still exists here in the year 3001. It is amazing how many problems have been solved. Cars are now run on water; their exhaust is pure water vapor. What a unique civilization they possess now."

In the helicopter Toni kept careful watch on John's face, waiting for signs of returning consciousness. His respiration was now proceeding unaided and his color was good. He looked as though he were merely asleep on the stretcher in the helicopter. She held onto his arm and prayed that he would come back undamaged by this experience. Suddenly John's eyes opened and he looked directly at Toni and smiled. Toni thought, He looks like a little boy with his secret smile, as if he had a secret and I were his mommy.

Inside, John, in opening his eyes, saw an aluminum plate over his head with rivets in it. He did not know where he was or when it was. The thoughts and the realities of the year 3001 maintained themselves as he opened his eyes. He did not know who Toni was and thought, Amazing, the help that one gets in the thirty-first century! He returned to his inside reality. He was traveling in a sort of closed container with an unknown power source. He did not know where he was going or why these people were taking him someplace.

The next time he opened his eyes, he was still in the inner reality of the future. What peculiar uniforms these people are wearing: short-sleeved, green uniforms. They seem very busy somehow working by new methods of medical help. I wonder what happened to me?

The helicopter landed near the hospital in the Valley, over the mountains from the house. Toni stayed with John while he was carried on the stretcher into the emergency ward of the hospital. While the emergency crew was taking care of John, she went to the telephone and called Burgess. She told Burgess that John had almost drowned in the pool and he said he would be over immediately. Toni needed Burgess's support at this point.

Toni told the attending doctor the story and mentioned that John had been taking K previous to this episode.

As John returned to the twentieth century, he realized he was in the emergency ward of a hospital and that the year 3001, although it had seemed very real, was now gone. He was once again back in his present body. He thought, This is going to be a rather difficult situation, explaining what happened without admitting to having taken K. I wonder what's causing that pain in my belly wall?

By that time John had been moved into a hospital room and was aware that Toni and Burgess were entering the room. He saw that Toni had been crying and that Burgess looked rather severe, as though containing himself and his comments about this latest episode. During the subsequent conversation he realized that both Toni and Burgess were holding back out of deference to his returning consciousness. He appreciated their diplomacy and did not pursue the discussion of what had happened.

John said, "I have an excruciating pain in my belly wall."

Toni went out to tell the doctor about this symptom. The doctor came in, did a physical examination, and prescribed a shot for the pain. During the next two days the pain kept recurring and was knocked out only briefly by the doctor's prescription. Neither the doctor nor John could figure out the cause of this pain. Later, John was to find that he had become allergic to K and that the pain was the result of the injection of it into the belly wall. Several weeks later he discovered that this pain was totally relieved by an antihistamine; however, no one at this point had thought of antihistamines.

Within a short time the doctor became very impatient with his M.D. patient, John. He went on a vacation and left orders that John was to be transferred to a private psychiatric hospital in the neighborhood.

One evening an ambulance was called and John and Toni went to the psychiatric hospital together. By this time the recent effects of the near drowning and the long-term effects of previous injections of K had worn off to the point where John began to function once again in his usual relation to the human consensus reality. As he was being signed in at the psychiatric hospital, he realized that medical judgments were being made about his mental competence. His clothes were taken away from him, he was given a hospital gown, and all his belongings were put under lock and key. If this continued much longer, he was going to have an irreversible social situation which might be very difficult to handle and might even require legal help. When someone arrived to take his history, he finally said to Toni, "I don't want to spend the night here. I don't care what that doctor said. I will not sign myself in and I do not want you to sign me in either."

Toni said, on the verge of tears, "Do you want to go home with me?"

John said, "Yes, I don't want to stay here. I want to be with you at home."

Toni went out and told the hospital personnel their decision.

The attendant came in and asked, "Do you want to sign yourself out?"

John said, "Yes. Definitely. And I would prefer not to have anything on the record about being signed in or being signed out."

The attendant said, "I don't know whether that can be arranged or not, but you can sign yourself out."

John's clothes and belongings were returned to him and he and Toni drove home over the mountains.

Toni: John, you must get help. K has taken over your life. Burgess and I feel that you should get help from Dr. Jolly W. I have talked to Jolly and he wants you to come to the Psychiatric Institute and sign yourself in for a few days and get rid of the K in your system."

"I don't want to go to any psychiatric ward. I can stay off K without any help. I don't want to generate any medical records in any hospital."

"John, you must go. I am reaching the end of my rope. You have almost killed yourself several times and I can't take the responsibility any further."

"OK, Toni. Without your help I must give up this exploration for now. Call Jolly and tell him I will come over."

John signed himself into the Psychiatric Institute and was assigned to a room on a locked ward. He was interviewed by various personnel who did a psychiatric evaluation of him. The pain at the site of the K injections began to recur. He was moved to a private room in the medical section of the hospital and the doctors attempted their diagnosis. He was given a thorough medical checkup, including X rays, and nothing was found.

Jolly came to see him, to bargain with him about K. After several long negotiations he agreed not to use it. The pain gradually subsided and he returned home.

John felt that he had not sufficiently explored all the parameters of K. He decided to do additional experiments on its long-term effects. For a period of three weeks he gave himself injections every hour of the twenty-four hours. He did most of this work covertly in the isolation facility. He was not aware that Toni knew what he was doing.

For a period of three weeks, he immersed himself in the inner realities created by K, projecting them onto his outer reality. He became convinced of the intervention in human affairs of the solid-state life forms elsewhere in the galaxy. He became convinced that it was necessary for him to go to the East Coast and warn the government about this extraterrestrial intervention in the affairs of Man.

He told Toni that he must go to the East Coast. She did not understand and tried to argue him out of it. He did not explain his mission to her. He felt that he could not share this mission with anyone but those in power.

He flew to New York and took up residence in a hotel near Central Park. He continued the administration of K to himself. He obtained further supplies of K in New York through the prescriptions of medical doctors who were old friends of his. The prescriptions were for migraine attacks.

He began to receive very strong messages to return to his old medical school, Dartmouth, at Hanover, New Hampshire. He flew from New York to a New Hampshire airport near Hanover. He arrived late at night.

He went into the men's room, gave himself a shot of K, came out of the men's room, and passed out on the floor.

Inside, he was in communication with the extraterrestrial solid-state civilization; outside, he was picked up by an emergency squad and taken to the Mary Hitchcock Hospital at Hanover. In the emergency ward he was recognized by a young psychiatrist, Michael. Michael had had personal experience with the use of K on himself and on other people. He had a conference with his chief of service. It was decided that he would accompany John back to New York.

By this time John had come out of the effects of K, recognized Michael, and agreed to go to New York in Michael's company.

Michael stayed in the same suite of rooms as John and watched while John continued his injections of K. John became reimmersed in the inner realities. At one point he continued his mission to inform those in power of the dangers of the solid-state propaganda on the human species. He called the White House and asked to speak to President Ford in Michael's presence.

A commanding voice at the other end of the line said, "What do you wish to speak to the President about?"

John: "I wish to speak to him about a danger to the human race involving atomic energy and computers."

The commanding voice said, "I will have to have more details than that. Who are you?"

John gave his name and continued his plea to talk to the President.

At this point Michael interrupted, said something into the phone to the person at the other end, and hung up. He then turned to John and said, "You are in very poor shape. According to my chief, if this happened I was to arrange for you to enter a psychiatric hospital here in New York."

John: "Michael, if you do that, you are my enemy."

Michael: "I am doing it for your own good and your future well-being. You insist on remaining under the influence of K and doing things unacceptable to other people, endangering not only your life and limb but your future reputation. If you will not agree to sign yourself in, I will obtain the help of two New York doctors who will commit you to a psychiatric hospital."

John: "Come on, Michael, you understand K and its effects from your own experience. But you do not understand its long-term effects and the benefits of lowering one's threshold for awareness of extraterrestrial influence. I refuse to sign myself in. You get out of here."

Michael: "I am a lot stronger than you are, and I can get the help of the New York Police Department if you won't cooperate. I have called two doctors who are willing to sign you in. I can't, since I don't have a license to practice in New York State, but these two doctors have agreed to do it. They will put you under the care of a third doctor who has an appointment at that hospital. He knows you and he also knows the effects of drugs on patients."

With these threats John gave in. The two doctors arrived, put him in their car, and took him to the hospital.

His clothes and the K that he was carrying with him were forcibly removed. He suddenly realized that he was being subjected to that disqualified, second-class-citizen, inferior status of the mental patient in a psychiatric-hospital setting. He remembered Thomas Szasz's writings entitled *Psychiatric Justice*. As the effects of K began to wear off, he realized that he was in a battle for his freedom.

He asked the personnel to contact the head of the hospital so that he could talk to him. The chief psychiatrist came to his room and discussed his case with him. He said, "The law allows us to keep you here only for a period of five days. If you wish to get out in spite of the attending physician, you must contact the Mental Health Commission of the State of New York."

John called the Mental Health Commission and talked to their legal counsel on the pay phone in the recreation room of his ward.

He called Toni long distance, explained what had happened, and asked for her help.

The attending psychiatrist came to see John. "John, you are obviously in a depression. I want you to take an antidepressant. Your use of K is a consequence of your depression. You can avoid the necessity of taking K in the future by using my prescription."

John: "I couldn't be less depressed. I have long since left periods of depression behind me."

Doctor: "You just don't understand your own condition. You have impaired thinking which can be corrected chemically. I insist that you stay here until you agree to take the antidepressant."

John thought, Chemical control of human beings. It reminds me of the old electrode story and the control of patients and animals. Tranquilizers, energizers, antidepressant drugs are the new repertory of mental control. The social machinery plus these drugs lead to control of people who are doing things that are unpleasant to other people. The virtuous nature of the justifications that those in charge have for administering the drugs does not make it any more acceptable. This doctor has no curiosity as to what's really going on in me. He refuses to spend time to find out what the truth is in me; he wants to impose control over my mood and my state of being through chemical means. I wonder how many of the drugs which he is prescribing to control people he has tried on himself to observe their effects, even as I have been doing with K?

During the next three days, the psychiatric personnel interviewed John, tested him, and found that there was no mental illness. By this time the effects of K had worn off completely. The head of the hospital reported to John that they could find no justification for keeping him, that he was being held only at the insistence of the attending physician who had been called in by Michael.

During a series of negotiations with the attending physician and the head of the hospital, it was arranged that John be transferred to another psychiatric hospital for further diagnosis. Meanwhile, the legal counsel of the Mental Health Commission had come and talked to John about the problem of getting out from under the commitment proceedings. She said that the commission could arrange for a hearing at the end of five days and free him from the hospital. However, this would mean a legal record and, hence, it would be preferable if other means could be worked out.

As a consequence of these considerations, John agreed to move to the other psychiatric institution, undergo further tests, and get opinions from those who were not involved in the original case.

Meanwhile, Toni had come to New York to enter into these negotiations and help John transfer to the other hospital. After interviewing John, the head of the second hospital said John was OK and agreed to take the responsibility for John's release, despite the disagreement of the attending physician. He realized that John had not called in that physician voluntarily, that he had been called in by Michael without John's consent.

John was released from the hospital and Toni took him back to California.

Despite the lessons learned in these near misses in regard to his life and his social freedom, John once more resumed his exploration of the parameters of K. He felt that his mission was incomplete and, driven by this mission, he continued the experiments upon himself.

Seduction by K

The year in which John was investigating the effects of K on himself, he had one overriding belief system or, more properly, metabelief system, which controlled his entry and exit to and from other belief systems. He called this overriding belief a "metabelief operator" (MBO). The MBO was: "In my development as a scientist I must approach the inner realities as well as the outer realities. I must investigate the properties of the observer/operator and his dependence upon the presence of changed molecular configurations within his own brain. K introduces certain specific changes in the molecular configuration and computation of that biocomputer. Some of these changes are visible to outside observers, some are visible only to the inside observer/operator.

"The scientific observer/operator exists within two sets of realities, *those of the human consensus reality, the external reality (e.r.), and the internal reality (i.r.). The e.r. and the i.r. exist simultaneously. The observer/operator exists in the i.r. sometimes interlocked with the e.r. and sometimes in isolation, not so interlocked. At high levels of concentration of K in the blood, the observer/operator is cut off from his interlock with the e.r., including the human consensus reality. The only physically safe and socially safe location to investigate this cutoff is floating in the isolation tank in a controlled environment, isolated from the necessity of interactions and transactions within the human consensus reality.* One of the dangers in this exploration is allowing this cutoff of interlock to occur outside the isolation facility. If in his exploration the observer/operator loses this perspective, he will inevitably be testing the limits of acceptance by the current human consensus reality, as an individual in the grips of a belief system counter to the current accepted belief systems."

During the first part of this year, John did experiments with single doses of K and arrived at a quantitative relation between the dose given and the resulting states of being induced in himself. Later he took multiple doses more frequently and found new effects not accountable simply by the induced phenomena of single doses at widely spaced intervals.

At the beginning of the year he did not realize the long-term effects of repeated doses. During that year he found that he entered into the overvaluation domain induced by repetitive doses of K. Toni called this "being seduced by K."

The first few months were taken up with the determination of the effects of single doses separated by several days. John worked in collaboration with Craig and several other young researchers. No one yet knew of the long-term repeated-use trap.

After the first dozen or so experiments designed to find the various thresholds for phenomena, John began to think in the following terms: "The time course of the effects of K after the time t_0, the time of the injection into the muscles: There is a very rapid movement through the various phenomena for the first few minutes. The effects then level off for a period of approximately ten minutes and then gradually subside on a much slower time course. These effects seem to be related to the changes in concentration of K within the bloodstream. If we think in terms of a time-concentration-in-the-blood curve (Figure 1K), we may be able to account for the results and the changes in the observer/operator and his belief systems during the half hour to forty-five minutes of each experiment.

"After the time of the injection, there is a period of about three minutes when no effects are felt. Rather abruptly, the effects begin and move rapidly through a series of phenomena too fast to be grasped. After this rapid rise of effects, there is a *stabilized plateau* where one experiences phenomena which are dependent on how much K one has injected. This phase lasts from ten to thirty minutes. By controlling the initial dose one can control the period of time over which changes in the self and in the i.r. are experienced. One can also vary the phenomena experienced on this plateau by the amount of K injected.

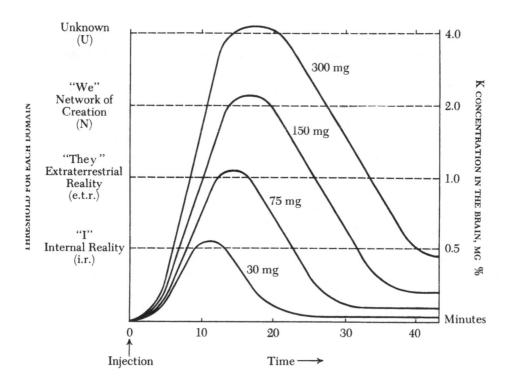

Fig. 1K: Dose Guidance Curves

"As the amount of K in the blood decreases because of its destruction by the body, one can then see and analyze the phenomena that occur over a period of twenty minutes to forty minutes.

"There are apparently no aftereffects that are detectable by the inside observer after the falling phase of K concentration in the blood." (Later John was to find that there was a small residual effect that lasted several hours. The falling curve did not go completely to zero. The overvaluation trap would be found much later to be caused by this small residual effect unnoticed in the first set of experiments.)

John did a series of experiments relating the amount injected to the phenomena experienced on the plateau.

He tried 10 milligrams at one injection. The effects were almost undetectable. There was a slight change in body sensations but no detectable change in himself.

He then tried 20 milligrams and found an enhanced body energy and tingling in the skin. There was no change in the visual field or in his perception of himself.

He tried 30 milligrams. After the initial rise of sensation, he began to sense changes in his perception (on the plateau). If he closed his eyes he could induce visual images: at first flat, two-dimensional, uncolored; and, a few minutes later, three-dimensional, colored, and moving. In this phase he became enthusiastic about the images, but not as enthusiastic as he had been under psychedelic agents.

He decided to test the difference between the various doses and the effects inside versus outside the tank. He started the tank work with a dose of 30 milligrams.

Freed from the effects of gravity, light, and sound in the tank, he was able to study the visual images in a more relaxed state. In the tank he saw continuous motion-picture-like sequences, highly colored, three dimensional, and consisting of, at first, inanimate scenes which later became populated with various strange and unusual creatures as well as human beings. He found that he could change the content of these internal movies by the self-metaprogramming methods he had learned in the tank and, in 1964, had used in the tank under LSD.

At this point he realized that if he stayed in the external reality outside the tank, these images became interlocked with that reality. They were modulated and modified by what was happening in the external world, whereas they were not in the tank because the e.r. was missing. There was some spontaneous source of these images in the tank as well as the modifications introduced by him as the observer/operator in the system. Early in the series he conceived of these spontaneous sources, i.e., something within his own brain which was generating the images in addition to his intentions for those images. At the beginning of this series of experiments, he assumed the existence of a contained mind, with the observer contained within that mind within the brain. Later he was to believe otherwise—that the source of the images was coming from somewhere else, not his own brain, by means which he did not yet understand.

He then went on and experimented with higher doses. He called the 30-milligram-dose threshold for visual projections the internal reality threshold, best seen in the isolation tank. The next amount injected was 75 milligrams. In the tank he found the plateau involved whole sets of phenomena which he had not seen at the lower doses.

For the first time he began to sense changes in himself other than the changes in perception of visual images. His relationship to his physical body became weakened and attenuated. He found that he began to participate in the scenes which were previously merely visual images, as if out there, outside of his body. His observer/operator was becoming disconnected from the physical body. Information from his bodily processes was becoming so weakened that there were times when he was not aware of his body at all. On this plateau he began to experience interaction with the strange presences, strange beings, and began to communicate with them.

"I have left my body floating in a tank on the planet Earth. This is a very strange and alien environment. It must be extraterrestrial, I have not been here before. I must be on some other planet in some civilization other than the one in which I was evolved. *I am in a peculiar state of high indifference. I am not*

involved in either fear or love. I am a highly neutral being,
watching and waiting.

"This is very strange. This planet is similar to Earth but the
colors are different. There is vegetation but it's a peculiar purple
color. There is a sun but it has a violet hue to it, not the familiar
orange of Earth's sun. I am in a beautiful meadow with distant,
extremely high mountains. Across the meadow I see creatures
approaching. They stand on their hind legs as if human. They are
a brilliant white and seem to be emitting light. Two of them
come near. I cannot make out their features. They are too brilliant for my present vision. They seem to be transmitting
thoughts and ideas directly to me. There is no sound. Automatically, what they think is translated into words that I can
understand."

First Being: "We welcome you once again in a form which
you have created. Your choice to come here, we applaud."

Second Being: "You have come alone. Why are you alone?"

I answer: "I do not know. There seems to be something
strange about this; the others are reluctant to join me here."

First Being: "What is it that you want from us?"

I say: "I want to know if you are real or merely a product of
my own wishes."

Second Being: "We are what you wish us to be, it is true.
You construct our form and the place in which we meet. These
constructions are the result of your present limitations. As to our
substance, whether 'real' in the accepted sense upon your planet
or 'illusion' in the accepted sense on your planet, is for you to
find out. You have written a book on human simulations of reality and of God.° Your problem here is whether or not you are
traveling in one of your own simulations or whether you have
contacted real Beings existing in other dimensions."

The scene begins to fade. John moves out of this *extra-*
terrestrial reality (e.t.r.), resumes his consciousness of his body,

°*Simulations of God: The Science of Belief.*

and sees the old familiar movies of Earthside scenes and his own memories. Slowly these projected images fade and John is floating in the tank, remembering them in full detail. He climbs out of the tank and dictates the experience into a tape recorder.

Thus did he find another threshold under the influence of K. He began to call this the *extraterrestrial reality threshold* on which his observer/operator became involved as a participant. The critical value of K at a single dose for exploring this realm was 75 milligrams.

The next threshold was found at 150 milligrams of K. In order to see this threshold clearly, he found that he also had to be in the isolation tank free of the interlock with the external world.

"I rapidly pass the i.r. threshold and the e.t.r. threshold, and suddenly 'I' as an individual disappears.

"We are creating all that which happens everywhere. We have become bored with the void. We know we have been eternally, are eternally, and will be eternally. We have created several universes, have dissolved them, and have created new ones. Each universe we have created has become more complex, more amusing to us. Our control of the current universe is on the upswing; it is becoming more complex as we regulate its regulation of itself. As we experience each universe, our awareness of ourselves increases. Each universe is a teaching machine for our awareness. To create a universe we first create light. We contain the light within the universe, within the space that we create to contain the light. We curve the space to contain the light.

"In the early universes we watched the light contained traveling through its empty spaces, bouncing off the periphery in the curvatures of the space. We played with the size of those universes, expanding and contracting them, and watched the light. Large universes finally bored us, the light merely traveled around and around.

"One universe that we created, we decreased in size until the light was chasing its own tail. We found a new phenomenon, a new effect. When we decreased the universe sufficiently, the light, in chasing its own tail at very small sizes, became stabilized. The universe became a single particle of incredibly small dimensions. The light, in chasing its own tail, had generated this particle which had mass, inertia.

"In the universe after that one, we created many small particles encapsulating light chasing its own tail in the small dimensions. We found that some of these particles attracted one another, forming larger assemblages. We played with these assemblages. We found that light within these particles, rotating in certain directions, caused the attraction of other particles in which light was rotating in the opposite sense.

"In a later universe we allowed the creation of huge numbers of these encapsulated light particles. We controlled their creation at one point and packed that small region with more and more particles. We found that there was a critical point at which they exploded outward.

"In a later universe we re-created the exploding point, and as the particles spread outward we arranged for them to condense on new centers. These new centers continued outward until we closed that universe and its space.

"In a later universe we began to reassemble particles in various parts of that universe, set up creative centers within the space of that universe. We set up points at which new particles were created and other points at which they were destroyed, reconverted into light.

"In a still later universe we allowed certain areas to become imbued with portions of our consciousness. We watched their evolution and found that each of these areas as it evolved became conscious of itself.

"In the current universe we have many assemblages of particles which have self-awareness. Some of them are huge, some of them are very small, a few have begun to question their own origins; a very, very few are becoming conscious of us. We are beginning games with these very, very few, manipulating their awareness. Most of these seem to be developing a sense of humor similar to ours. This universe is more amusing than the past ones."

John's consciousness and self-awareness condensed back into a single individual. He began to experience himself as a self, separate. He came back through the e.t.r., into the i.r., and finally into his body in the e.r., which was the tank. He labeled the domain of losing self and becoming "We" the *Network of Creation (N)*.

He then tried the 300-milligram-threshold dose. He found that this plateau was beyond anything he could describe. It was as if he had entered a void, had become the void beyond any human specification. In returning from the void, he went through the creative network, the extraterrestrial reality, the internal reality, back into his body in the tank. He realized that, as a human being, he would be unable to use these larger-dose regions. He would be unable to describe what happened, so he labeled this high-dose threshold *U, the Unknown*. At this point he abandoned study of the higher doses leading to the Unknown (U).

He now began to see the dimensions of the exploration and the parameters he had to explore. He divided the experiments into those to be done in the tank and those to be done in the human consensus reality, with single other individuals and with unprotected situations, not in his home. New dangers were to appear, one of which would terminate this exploration and render him incapacitated in a bed at home for a period of twelve weeks.

K and Interlock
with the
External Reality

For twenty minutes Craig jumped up and down on spread feet, bent knees, and body bent at the waist with his arms curled forward, howling like a chimpanzee. As the howling, the motions, and the posture returned to that of a human, John said to Craig, "Where did you go?"

Craig: "I went back to the beginnings of the evolution of Man. I became the predecessor of a caveman. I saw a saber-toothed tiger and howled my defiance at him. As he left I ran for a tree. I climbed it and, as I was coming out of K, I found myself sitting in the tree looking at you."

John: "Do you want to know what you looked like out here?"

Craig: "Yes."

John: "You looked like an epileptic having a petit mal seizure. You were staring straight ahead and acting as if you were a chimpanzee howling. I will play back the tape and let you hear how you were sounding for that period of time."

As the sounds of Craig's cries came from the tape recorder, his face lit up and he began to laugh.

John: "You can afford to laugh. I had to stand here and listen to that and watch you, hoping that you weren't going to attack me. I would like to make a strong suggestion to you. If you ever do that again I'm going to kick you in the rear end."

Craig: "Ah, come on, Doc, don't take it so seriously. Let's find out what we can about this. It looks as if automatisms can take place under the influence of K. Just before I went under, I decided on the regression program back to the beginnings of Man. It was a very interesting domain which I've explored before with other chemical agents in me. Apparently we have primitive primate programs buried in our central nervous system. We can activate these in special states of being. The question is whether we are creating these from our own fantasies or whether they are really built into the brain. This is a fascinating puzzle."

Craig had injected 100 milligrams of K before this experience started. John and Craig agreed that he had passed through the i.r. threshold and the e.t.r. threshold but had stopped short of the network threshold. This decision was based on the fact that he maintained an individuality, that he had not been immersed in the creative network, becoming a "We" rather than an "I." Craig described his surroundings during the experience as the surface of the planet Earth something on the order of six million years ago. In the experience he possessed no human language. His postexperience judgment was that he was an organism of the primate series which had preceded the evolution of humans as we know them.

The series of joint experiments then continued with Craig in the tank and John as safety man. Thus they discovered the most dangerous aspect of K exploration.

As both Craig and John began to realize, during the period from the injection to the immersion in the internal reality (i.r.) at the higher dosages, it was imperative that there be controls on the body. The safest place to make these transitions was in the tank in a preprogramming leading to bodily immobility. The excitation of motor mechanisms within the brain tended to persist beyond the point at which the observer could control them. The brain "went on automatic" and the observer lost control of it under these particular circumstances. They agreed that there was a dangerous disengagement of the self from the bodily control systems, and the bodily control systems persisted in actions which they were taking before this disengagement took place.

In Craig and John's mutual explorations, they arrived at for-mulation of the first danger of K: *At certain critical doses and certain critical concentrations of K in the brain, the subcortical systems continue their automatic activities out of contact with the observer in the brain. Do not ever get caught without a safety man when exploring this domain. Restrain such experiments to the isola-tion tank with a safety man present.* (In later years, reviewing the happenings of that crucial year, John came to call these directions he and Craig had worked out the First Prime Directive.)

John now entered a new phase of investigation of K and its effects on him. He and Craig became temporarily separated and John pursued the work on his own.

He began his exploration of the effect of small repeated doses of K. He wanted to find out if by means of such doses he could remain above a given threshold (see Figure 1K) for longer periods of time than he could achieve with a single isolated dose. This experimentation was limited by his ability to give himself a dose. If a previous dose put him above the extraterrestrial reality (e.t.r.) level, he found he was unable to carry out the operation of giving himself an injection.

In the first experiment in this series, he began by giving himself a 30-milligram dose. He found that he just touched the internal reality (i.r.) threshold on the plateau of the effect. He repeated the 30-milligram dose at fifteen minutes from the first injection. On the plateau, he found that he had moved above the i.r. threshold and began to see visual movies with his eyes closed. After fifteen minutes he gave himself the third injection and saw that at the peak he was above the e.t.r. threshold. He let the effects of K wear off for twenty-four hours.

He had found a new effect of K. Doses were additive if timed correctly.

The next experiment involved repeated doses of 75 milli-grams (Figure 2K). On the first dose he passed the i.r. threshold and hit the e.t.r. threshold. He was immersed in a scene in which he was participating which was not connected to his external reality. The scene had the alien qualities characteristic of the e.t.r. threshold. As he returned from this participation into the i.r. region and finally back to the external reality, he gave himself the second dose of 75 milligrams.

This time he passed the i.r. threshold and went deeply into the e.t.r. domain, farther than he had gone with the first dose. For a few minutes of clock time he participated in alien regions beyond his comprehension.

As the K effect began to wear off, he came back to the e.r., floating in the tank with the door open and the lights on in the room. Almost automatically he looked at the clock, saw that fifteen minutes had passed since the last dose, and gave himself the third injection.

This time he rapidly penetrated through the i.r. and the e.t.r. and joined the network (N). As he was split out from the network, he traversed another alien extraterrestrial environment

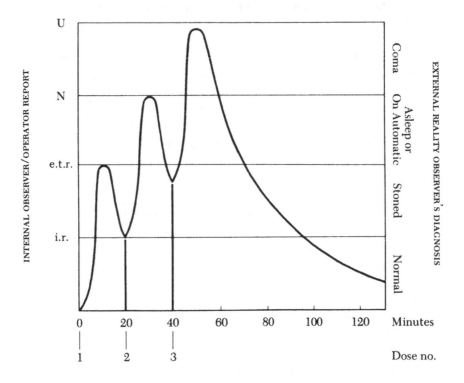

Fig. 2K: Multiple Dose Guidance Curve

(e.t.r.), became immersed in his own internal reality (i.r.), in which he began to remember having given himself three injections, and came back into his body, the tank with the door open, and the lighted room.

As he lay in the tank contemplating these results, he became very enthusiastic and quite exultant about the phenomena experienced. He now had a certain measure of limited control over traversing into various regions for short periods of time.

He repeated this experiment once a day for the following week. He began to detect a new effect.

He found that there was a carry-over effect from the multiple doses of one day, enabling him to move more easily into the inner spaces, the inner domains, more easily with each experiment. By this time his motivation was changing. He began to value these inner experiences more than he did his relationships with the external world and other persons. He began to take on a new belief system.

"Somehow I am being programmed by sources far greater than the human. In my old terminology, super-self metaprogramming is taking place. I am doing these experiments at the behest of someone or something far greater than I or the human species. Whoever they are, they want to do these experiments.

"They seem to be the two Beings that I met early in the series, some sort of teachers or guides who are insisting that I pursue this particular mission."

About this time Toni became aware of changes beginning in John. She found that he was spending less and less time with her and more time in the isolation facility on his experiments. He was becoming less available for their dyadic joint life together.

Meanwhile John was having difficulties obtaining sufficient K. He had several sources; one of these became aware of the fairly large amounts that he was requesting. This source refused to give him any more. He found other sources and continued the work.

The new belief system of being programmed by suprahuman entities became his primary motivation. He had given himself so many injections that the pattern of giving the injections had become automatic. While still immersed in the i.r., he could give himself an additional dose to penetrate into the deeper regions. The injection process had become an automatism no longer needing his conscious control of what the body did. Long ago he had abandoned his and Craig's First Prime Directive.

He entered into a new, far more dangerous phase of his exploration and his new mission.

Living in
the Internal Reality
and the Extraterrestrial
Reality

21

John decided to try to live in the internal reality, continuously for an extended period, peaking into the extraterrestrial reality (e.t.r.) and contacting the network (N), staying out of the Unknown (U).

He found, by giving himself 50 milligrams every hour on the hour, twenty hours a day, with four hours out for sleep, that he was able to maintain the schedule for three weeks. The experiment ended at three weeks as the result of an accident in the external reality.

By the end of the first day he was able to maintain the i.r. because of the accumulated effect of the many doses that he had taken. No matter what he was doing, he could close his eyes and see colored three-dimensional motion pictures. He studied these in the darkened tank for many hours at a time. He found that they would be maintained if he sat in a bathtub in a dark bathroom. Even though his activities in the external world were restricted, the i.r. came to be present continuously, day and night.

After each dose of 50 milligrams every hour on the hour, he spent some time in the extraterrestrial reality. At the beginning this time was approximately twenty minutes. After the first week it had lengthened to a period of forty minutes. During the second and third weeks, he felt the extraterrestrial reality continuously, even when the internal reality and the external reality were available. The boundaries between the three realities became less definite. He found that while he was moving around in the external world, he could feel the influences of the Beings of the extraterrestrial reality (e.t.r.); any time he closed his eyes, even in bright sunlight, he could see the internal motion pictures.

"That man I see on television is a direct agent of the extraterrestrial reality controlling all human life. He is giving a public speech on television to the human species in order to program them into believing that he is not an extraterrestrial agent. In reality, he is controlled by the solid-state life forms of the civilization of another place in our galaxy. It is obvious that what he is saying is to hide his real mission."

John was watching and listening to a powerful figure in the United States government, Elliot Richardson. As he spoke the television set became brighter and there was a sense of importance to what he was saying. When someone else came on to make comments about what he had just said, the set went dimmer, the picture was less clear, less important.

John thought, "The solid-state systems on Earth are being modulated by solid-state life forms elsewhere in the galaxy. This television set and all the communication links hooking it into the television networks in the United States are being controlled by a solid-state entity elsewhere."

Suddenly the channel he was watching went blank; the TV station was cut off. It stayed off for one minute and came back on again.

John thought, "The solid-state entity is demonstrating its control to me by this technique."

During these three weeks Toni took John to see the movie entitled *Executive Action,* based on a conspiracy to assassinate John Kennedy. Watching this movie he saw the same effect that he had seen on the TV screen. Certain portions of the movie were enhanced in brilliance and took on a heightened significance.

John thought, "I am being controlled by the solid-state entity. The control of my brain by this entity causes portions of this film to be emphasized to teach me those portions which are messages from the solid-state entity."

During this period John became convinced that he was a visitor from the year 3001. He felt that everything around him was happening in that year. When traveling in automobiles he felt the primitiveness of their mechanisms. He longed to return to the year 3001 in which the vehicles produced steam instead of contaminating exhaust gases. Everything he saw had a patina of old age, as if in a museum, including human beings. Clothing and machines seemed old, antique. Paintings looked as though they had been made by artists long since dead. He briefly began a search for the time machine which had transported him backward in time to the twentieth century. He could not find it and realized that the Beings from another dimension had this control over him. He developed the belief that a Being from the year 3001 had taken over his body. After a few days this Being left and he returned to his twentieth-century self.

While John was believing that the Being from the year 3001 was literally himself, he interacted with the people in his external environment in ways that were totally uncharacteristic of him before his seduction by K. To them he had become a very young, very naïve boy. He did not interact in their conversations in his usual quick and witty fashion. Instead, he apparently was immersed within himself, yet watching them and their interactions. Every so often he would perform some out-of-context action which resulted in their surprise or shock.

Once, a friend who had interviewed John for a magazine visited the house with his new girl friend. The young boy who was John became entranced with her beauty.

He stared at her beautiful face unremittingly. He then had the impulse to fondle her breasts, walked around behind her, put his hands on her breasts, and fondled them. He ignored the shocked reactions of Toni, of his friend, and, of course, of the young woman, and immersed himself in a fantasy with her. In his fantasy she and he were making love as two naïve children would make love.

The man involved in this episode was a protagonist of the

new-consciousness, human-potential movement. He could not accept the obviously changed John and from that point on maintained a safe distance from him and Toni. Even several years later he could not resume his old friendship with John.

Toward the end of this three-week period, John noticed a broken shelf beside his bed. He asked Toni how it had gotten broken.

She looked at him rather quizzically and said, "You fell on it last night."

John couldn't remember this accident. This brought him up rather short and he realized that dangerous automatisms and coma had entered into the picture. He stopped taking K. He found that it took him about three days to lose the internal reality effect and the extraterrestrial reality effect.

He went into isolation and tried to work out what had happened during those three weeks. He began to realize that the old picture which he and Craig had constructed had become very different. The sharp thresholds which they had postulated for single doses did not apply to chronic hourly administration of K. These sharp thresholds during the three weeks had become coextensive. The external reality (e.r.), the internal reality (i.r.), the extraterrestrial reality (e.t.r.), and at times even the network (N) had grown very, very close together. During the chronic administration, a single dose would re-create all of these phenomenological domains. Between doses the belief systems inherent in these domains took over all his relationships within the external human consensus reality. During the three weeks, he had taken some five hundred doses of K. Not only was there not enough time between injections for the K in his body to be destroyed, but the whole structure of his belief systems was changed, lowering the thresholds for the effects of experiences in the i.r., the e.t.r., and the N.

Subsequent to the three weeks of experimentation, John and Toni gave a workshop together. They had contracted to do it before the period of K. There was some doubt in Toni's mind as to whether John could function at this point in his usual way, but they drove to Big Sur and started the workshop anyway. John continued the injections of K during his exposure to participants of the workshop and members of the staff of the institute.

He felt as if he were an Indian sage or an extraterrestrial being who was observing a group of humans on a peculiar planet called Earth. To Toni, those in the workshop, and the staff, he seemed very withdrawn, extremely calm, as if he were continuously in the state called Samadhi. In exchanges with members of the workshop, he observed and participated only rarely.

At one point he listened very carefully to a man from Texas who kept repeating himself as he had done in previous workshops. Even though he had been living at the institute for long periods of time, he obviously had not made much progress in changing himself. He protested about his difficulties and at the same time kept talking about his Cadillac car.

In John's internal reality this man was merely repeating an old pattern and not really examining himself. Despite his protestations that he wanted to change his states of consciousness, he was not doing anything to accomplish this. He asked John what he should do.

John replied from his position of removed, objective, extraterrestrial sage, "I hear you driving your Cadillac. I do not hear you driving yourself to change yourself."

The man reacted by withdrawing within himself, shocked by this direct challenge.

In the presence of this strange, extraterrestrial John, Toni took on the burden of replying to and helping to program the members of the workshop. Outside of the sessions she was given considerable support by the staff of the institute.

Toni learned who could be tolerant of changes in states of being and states of consciousness and who could not. She watched the interactions of staff and others with John and realized that there were very few persons who could perceive such changes, accept them, and interact with John in an effective fashion. To those few, his behavior was completely understandable; to the others it was totally outrageous. She saw that the human world divided into those who had real insight into states of being and consciousness and those who only said that they had such insight.

By this time all his sources of supply of K had become wary. Some of the friends with whom he had left K became very cautious and refused to give him what was in their care. He went through a severe withdrawal period in which he missed K and its induced effects. During this period he was described by one of his best friends to be "like an alcoholic who could not obtain alcohol." John vehemently denied this parallel, still caught in the seduction of K and still caught in his mission.

He decided to try a chemical substance closely related to K but which had longer-lasting effects. He had not tried this substance, but he had heard a good deal about it from friends among the young M.D.s who had used it. This was twelve months after the beginning of the experiments with Craig. Up to this time Toni had accepted what John was doing and justified it to the many friends who objected. She had stood by John in spite of his remoteness, in spite of his isolation, and in spite of his separation from her inside his own head. She knew that she was married to an unusual, unique human individual who had purposes very different from those of anyone else that she had ever known. At one point she said to a friend, "If he were an ordinary person doing these stupid things, it would be easy for me. I could pick up and leave. But he is not. I can't expect to lead an ordinary, quiet life while living with a genius."

Much later an interviewer asked Toni what it was like to live with John. She said, "Sometimes I yearn for trivia."

John obtained the substance related to K, kept it hidden, and waited until Toni left the house. After she was gone he gave himself an injection. John didn't know that it required longer to take effect than K. He made a strategic error in not realizing the longtime course of the buildup, the plateau, and the decay of this new substance in his brain. He had not prepared himself to stay in the isolation facility and in the tank for the requisite number of hours. He made the mistake of continuing to walk around in the external reality.

As the new substance began to take effect, John was transported into the joys of his childhood. He became a little boy once more, happily enjoying the sunshine, the flowers, and the plants around his home. He felt his body and his motions with intense, rewarding pleasure.

Under the influence of the new, longer-lasting chemical similar to K, he responded to a call for help from Toni.

Several days before, John had installed a locking gas cap on the VW camper. Toni had left the house and taken the VW down the canyon. At the nearest gas station she tried to get gasoline, found the gas cap locked, and realized that John had not given her the key. In response to her call, John decided to meet her down the road to give her the key.

John got on the bicycle, looked down at his feet with great joy, and said to himself, as it were, "This is the first time I've been on a bicycle with such joy since I was a little boy first learning how to ride." John went down the road on the bicycle, met Toni, and gave her the key. She turned the VW around and went down the canyon road. John followed her.

In the state induced by the chemical, his interlock with the external reality became disconnected. The bicycle was traveling about thirty miles an hour down the twisting, turning road. Suddenly the chain came off, the rear wheel locked, and John hit the road. He tried to roll to avoid damage but landed on his right shoulder, breaking his collar bone, his scapula, and three ribs and puncturing his right lung.

Third Conference
of Three Beings

Third Being: "I have called this meeting to report the close of the year of education of my agent on the planet Earth. I would like to give a short review of that year and its termination.

"I and several other Beings with agents on the planet Earth were very busy at this time with the coincidence control of this particular agent. I operated in close cooperation with the Second Being, who was controlling the agent called Toni.

"There were times when our coincidence controls had to be introduced with an exquisite sense of timing. At one point the Second Being and I arranged for the agent Toni to read an account of mouth-to-mouth resuscitation, preprogramming her for an episode that happened three days later in which she found the body of the agent, John, floating in the swimming pool. Another Being had to be enlisted to time the publication of that account of mouth-to-mouth resuscitation in the paper which Toni sometimes read. Coincidences were arranged for her to buy that issue of that paper and read it.

"In this episode we enlisted the aid of another Being who prodded another agent, Phil, at the critical time to make a telephone call trying to reach John through Toni. This led her to try to find John and to discover his body in the swimming pool. The previously mentioned coincidence control pattern allowed Toni to save him by mouth-to-mouth techniques."

First Being: "I note that the educational process was terminated. How was this accomplished?"

Third Being: "This was done in cooperation with many agents, who gradually arranged for coincidences to cut off the supply of the substance K which John was using during the educational process. The human vehicle had become 'seduced by K.' By this time he was sufficiently educated to seek a longer-acting substance similar to K but lasting several days instead of minutes. By arranging a series of coincidences, I led John to obtain a large enough amount of this substance for the final scene of the termination process. I arranged for him to purchase a bicycle and use it at the critical time of the oncoming effects of that substance, and with the cooperation of the Second Being I arranged for Toni to leave the scene.

"At this point exquisite, fast timing of events was necessary.

"As John was riding the bicycle down the canyon road, I chose a place for the chain to lock the rear wheel. At this particular instant his body was in such a position that as he fell he received painful but not fatal injuries. I arranged for his biocomputer to be uninjured to preserve the vehicle for a future mission."

Second Being: "It took a good deal of careful preparation over a long period of time of the agent, Toni. Her protective care of John had to be preserved for future use. And yet it was necessary to remove her from the scene so that we could terminate his use of these substances. There was a delicate balance of maintaining his cover of the use of the chemicals and, at the same time, arranging the supply of one more substance in the human consensus reality and inducing the agent Toni to leave so that the process could be completed. Later she experienced some guilt about being away from him at this particularly crucial time; however, her education as an agent on Earth required this series of coincidence controls."

Third Being: "In cooperation with several other agents, we overdetermined the coincidences so that the agent John would get to the hospital with no further injuries. We arranged for three persons to call the Rescue Squad, the sheriff, and the Fire

Department. Once the agent John was injured, we made sure he would be taken care of in the human consensus reality despite his inability to take care of himself as a result of the injuries and the substance within his biocomputer.

"It was realized that he must be put through a new phase of his education while in the hospital. While he was in coma, sedated, tranquilized, and under the influence of painkillers, the Second Being and I took him to other planets on which catastrophic planetwide disasters were taking place. He wrote these experiences up later in a book."°

Second Being: "It was necessary to arrange for the education of the agent Toni during this period. Throughout John's educational process, she was educated in parallel. I cooperated with the Third Being throughout that year. In the hospital I arranged for her to test John Junior, the son of John, and to educate him further as to who she really is. During that year the agent Toni learned that she can depend on her deep self; she learned that she has deep reservoirs of loyalty, of trust, of strength which she can exert on behalf of her own human vehicle, on behalf of John, and on behalf of her extended family. She learned all this despite the fact that it was a very stress-filled period for her vehicle."

Third Being: "By this process the coincidences were arranged to put the agent John through a twelve-week period of intense physical pain. As a result of the accident, he refused the use of any further chemical substances, including those prescribed by his doctor. We arranged for him to be twenty-four hours a day in his home with Toni. He was given time to integrate the previous year, to integrate the essentially delicate and mortal nature of his physical body and its recuperative powers. This training freed him of the necessity for using any further chemical agents. He had found that his biocomputer tended to become simplified in its programming under the influence of K. He made many discoveries about the relation of the self to the body, to the external world, and to the human consensus reality. The reverberations of this educational year of John's are still re-

°*The Dyadic Cyclone,* pages 146-48 and 153-54.

sounding in the human consensus reality. Only at the present time have I felt that he can share his education with other humans who need the knowledge he has acquired. Other human agents now need to know his story.

"At this meeting I would like it to be decided whether this story is to be released as he is currently writing it."

First Being: "Let us decide that question later. I have been told by higher levels that during that year he was allowed to penetrate above our level. What do you have to say about this?"

Third Being: "Yes, several times I accompanied him into what he calls the extraterrestrial reality, and he saw and communicated with his simulation of the Second Being and myself. During that time we communicated directly with him. When we felt that he was sufficiently educated about us and our abilities to take care of him in the other domains, we arranged for him to penetrate to the higher levels with us. He and we became dissolved in the way familiar to us into what he later called the Network of Creation. He and we lost our identities and fused with those whom he calls We.

"After a sufficient immersion in the Network, we returned him to his body on the planet Earth."

First Being: "As I remember, several years ago you transported him to this domain several times."

Third Being: "Yes, we did. At that time he wasn't quite ready to accept the existence of the higher levels, and he later attributed the experiences to his own simulations of the creative process in the universe."

First Being: "What is the current status of his belief about us and about the Network of Creation?"

Third Being: "He tends to oscillate somewhat between belief in the reality of our existence and belief that these experiences are a consequence of holding to our existence as if real. There is still somewhat of a tendency to flip from belief in us as real to belief that this is a simulation."

First Being: "I wish to defer to a later meeting the question of whether he is to be allowed to publish the current account of his education and to break cover on what he has found out. The long-term coincidence control pattern may include such publication. At this time we estimate from information derived from higher levels and intelligences which both of you have collected that the probability of publication is not yet certain. The decision is not completely in our hands; higher levels insist on further evaluation and correlation with other agents on the planet Earth. I will let you both know at a later time how these probabilities are developing, and at the next meeting we can make the decision whether, at our level of control, he is to be allowed to publish.

"This conference is hereby declared adjourned."

Life
Refuses Closure

Are the three Beings real? Are any not-human Beings real? Am I being advised by them? Will they arrange for coincidence control to publish what I have written? Will there be repercussions from the human consensus reality if what I have written is published? Can I obtain the advice of the Being in control of my coincidences?

Despite sixty-one trips about the sun on the spaceship Earth, my body still functions more or less satisfactorily. It still disturbs me at times. It teaches its rules of maintenance and repair whether I like them or not.

Is my mind merely the computations of my brain? Is there something in me extending beyond me, as has so often been promised by my inner realities? When this human vehicle dies, will there be something of me and beyond me which continues? Is my affirmative answer, "Yes, Beings beyond humans exist," merely a product of the desires of my brain and my body to continue to survive beyond their earthly existence?

Is all of me—my consciousness, my awareness, my thinking, my love, my relations with others—terminable or interminable?

Man's fondest illusions are packaged in me even as they are in all others. My programs for survival seem to be built into the genetic code assuring the future of the species. Somehow, once protoplasm formed on this planet, it exerted a consistent inner force to reproduce itself and continue inhabiting the surface of the earth.

Is our love and compassion merely an expression of this innate characteristic of our protoplasmic origins? Are we more than individuals of one hundred billion cells each, totaling the four billion human beings on this planet?

First Being: "The higher levels of coincidence control have instructed me to call this meeting. Long-term human coincidence control patterns on the planet Earth are changing, pointing to a critical, crucial decision for all life on that planet. The human race now has the power to decide whether to continue life on earth or wipe it out in any of five different ways. The first method currently would be to totally contaminate the surface of the planet with radioactivity and nuclear explosions. They are playing games with *nuclear weapons and nuclear power*. As we know, other planets have killed off all life as the humans know it by nuclear methods.

"The second destructive force available is *chemical and biological agents* inimical to all mammalian life forms including the human species. Individual human agents have been killed with these substances. Nerve gases and strange new toxins and viruses have been created artificially.

"The third method involves their beginning understanding of their own structure. They have discovered those molecular configurations of which their own genes are built. They have begun to see that they are not too different from other animals or even from plants. Their new understanding of this basic building block of themselves and of all life is leading them to experiment with *artificial forms of life*. Since they do not understand as yet the total interdependence of all organisms on that planet, they do not realize that millions of types of organisms have disappeared because their genetic code was inappropriate for the particular time, place, and climate in which they developed spontaneously. Without such knowledge they may produce new organisms, new to them, which will be distinctly out of place, out of time and out of the evolutionary sequence established on earth. They may discover some organisms which can eliminate them and all other mammals.

"The fourth danger area in which they are caught has to do with the organization of *large human groups with belief systems counter to survival* of themselves and other human groups. There are peculiar human values connected to territory, connected to such dogma as, 'My beliefs are greater than yours; you must take on my beliefs or I will kill you.' They have developed methods of training their children in their traditional beliefs. These entrenched beliefs then cause the children as adults to cohere to one another and to their traditions in ever larger groups. Such human structures lead them to use other methods on one another in order to kill rather than educate those they call their enemies. As their power over natural processes increases, these dedicated groups are likely to obtain enough control over the first three methods to sacrifice the rest of humanity and the other organisms of Earth in a futile struggle to master one another.

"The fifth area of potential destruction is the *human species' arrogance*. Humans have a pride in their knowledge, as if this knowledge were complete and possessed closure. Only very few humans realize that pride in one's knowledge and shame of one's ignorance lead to the arrogance which is threatening the other species on earth. The humans continue to kill other species, irrespective of the large, fine brains which those species possess. The large minds of the whales and the dolphins are not granted credence by Man. The large, fine brains and large minds of the elephants are totally ignored; these animals are killed and controlled by groups of the smaller-brained, smaller-minded, ignorant humans.

"Have either of you found any signs on the planet Earth that long-term coincidence control patterns should be modified to include the survival of water-based life on that planet? Do you have any information that I can carry back to the higher levels of coincidence control which determine long-term coincidences to forestall the demise of the consciously aware organisms on the planet?"

Second Being: "As you know, the human consensus reality has been mainly male dominated. The reproduction of the species has been left to automatic programs within the males and the females. Most male humans are not aware of the automatic nature of their drives to reproduce. Most female humans have been forced by their long pregnancies to devote more of their thought, their feelings, and their relationships to that which carries the human species forward. Within each reproductive female there is something which understands and intuitively feels what will lead to survival of all species. However, the females are subject to the restless, aggressive programming of the males. Currently they are beginning to organize themselves, and those males who can hear them, into groups which value continued life above tradition. Females who develop confidence in their intuitive understanding of life, such as my human agent, Toni, are becoming confident of their powers to exert their influence in the human consensus reality. They hope to prevent the disasters brought about by the predominantly male scientific advancements."

Third Being: "I have some data derived from my relations with the human agent on Earth, John. Our education of him has led to his abandoning further experiments such as those he did with brain electrodes, with LSD, and with substance K and the isolation tank. Over his life we have finally educated him to the point where he can see much more clearly the dangers to all species of the present science, engineering, and social structures. We have subjected him to enough ruthless education in what he believes controls the universe. He realizes that his past illusions have been fed to him through his own traditions. Our training course subjected him to seduction by female beauty, by chemical substances, by scientific knowledge, by spiritual development methods, by power and money, and by death wishes; and finally came the seduction of him by the desire to become one of us.

"Currently he is involved in integrating himself as a human agent within a male–female dyad. He is beginning to see that he must play a role in the education of humans as to the future necessities of life upon the land and in the oceans of Earth. He is now willing to drop his consideration of his past life as a closed phase of his own evolution."

First Being: "I hear from other Beings controlling other

agents on the planet Earth that there are new currents of feeling
and compassion among large groups of the younger generation.
These agents also report that solid-state life forms have established
controls among the human reality. They have entrained hundreds
of thousands of humans in their mission to establish a solid-state
life form to take over the planet Earth. The problem of the higher
levels of coincidence control is whether the solid-state life form
should be allowed to be developed by the humans."

Third Being: "This is a probability that many humans are
aware of. They are insisting that the solid-state devices which
they have developed, their computers, are to be programmed in
the future only for the survival of all organisms upon the earth,
including the human species. Use of computers as weapons of
war is to be banned. Those who understand this are gradually
gaining a foothold in the human consensus reality."

First Being: "From your experience, does each of you feel
that there are sufficient numbers of humans who will educate the
rest and spread the knowledge of these dangers and the knowledge of alternative paths for humans to take?"

Second Being: "Among the women of Earth there seem to
be enough individuals to begin the movement toward the intuitive feel for life and its preservation."

Third Being: "Among the men of Earth there is new scientific knowledge which they realize they must be responsible for.
The implications for the end of human civilization are now understood in greater detail and programs have arisen with hundred of thousands of participants for the conservation of life on
the planet."

First Being: "I understand, then, that both of you feel life on
Earth has a chance of continuing. There are enough humans who
refuse to allow life to have a closure so that you feel the long-
term coincidence control patterns should be arranged for at least
a few more years for the humans on the planet Earth. I will take
that word back to the higher levels. We will see if such a future
can be arranged for Earth and its organisms."

Third Being: "I will now take charge of the coincidences for
my agent. I will arrange for him to travel to a meeting at which he

will once again contact the dolphins and whales. He will also be educated in the attitudes of other humans to the Beings known on that planet as the Cetacea. There are a few human agents whom he must contact in pursuit of that part of his education."

First Being: "We should refer at this juncture to the other Beings in control of coincidences. The time is arriving when there should be a demonstration to your agent of the essential mortality of his human vehicle and the chances he is taking with it. We must contact other Beings in control of other agents and set up the coincidence-control pattern for the next fourteen trips of the planet Earth around the sun."

John ruminated about the dolphins, the whales, and the porpoises. He thought of new devices to be constructed of solid-state parts which would be necessary to open a technical doorway to the Cetacea. He realized that, in order to convince Man of the high level of intelligence, compassion, and sentience of the Cetacea, the interchange of ideas between humans and Cetacea must be promoted. The fifty-million-year accumulation of wisdom by the whales and dolphins, their knowledge of the total interdependence of organisms of the sea, must be opened up for Man.

He conceived of new ways to convince other humans of the large minds of the Cetacea. He realized that computers must be used to train dolphins and whales to communicate with Man in those areas where Man feels preeminent—his computations, his logic, his consciousness, his relationships, his communication, his truth. The dolphins and whales must have teaching programs giving them codes with which to operate the computers to solve problems posed by Man—in computations, in logic, in self-initiative. One demonstration of this was badly needed. He hoped that within the next few years he would be instrumental in bringing about this beginning of a new appreciation of the capabilities of life in the sea by Man.

He had a new understanding of his skepticism about his own ideas, about his own beliefs; he hoped it would help him to apply himself to his new tasks. He felt he must rise above his pessimism and enter into the dolphin puzzle with his full powers, understanding, interest, dedication, and whatever financial means he could find.

He began his task by organizing, with his wife, Toni, Burgess Meredith, Victor di Suvero, and Tom Wilkes, the Human/ Dolphin Foundation in Malibu, California. He began to make contacts with those who could help in the financing of the new projects. He sought those younger engineers, scientists, musicians, artists, biologists, and others who could accomplish the mission. He realized that he did not have much time left on the planet as a human being. He also realized that his influence was relatively limited. It would be necessary for him to use all accesses available to him within the human consensus reality in order to carry out the desired program.

He saw that he must write an authoritative book about the mission of communicating with the Cetacea. Such a book was needed in order to provide the facts to those who wished them, insofar as these facts had been determined to date. The book also might function as a textbook to educate future generations about their kindly neighbors in the ocean.

JANUS Project

Steve Castillo

Jennifer Yankee at the control panels working with Tom Fitz training Joe during the JANUS project.

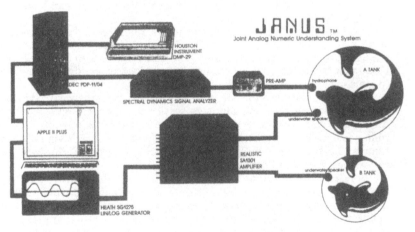

Dawn Rodia

JANUS (Joint Analog Numeric Understanding System) Project

Toward a Fluid Dialogue. Designed by the Human/Dolphin Foundation, JANUS allows researchers and educated dolphins Joe and Rosie to send and receive whistle-coded communiqués.

In a 1979 Pasadena conference, John Lilly talked of equipping a GMC stepvan with computers, signal analysis equipment, and software to create a dolphin whistle language. The intermediary language that evolved was named after the god of portals, Janus; whose initials stood for Joint Analog Numerical Understanding System.

A team of us formed at Marine World Africa USA in Redwood City, around John Lilly and Project JANUS. We all worked as volunteers, because we loved being near the dolphins, Joe and Rosalie.

Joe and Rosie were two Atlantic bottlenose dolphins captured for this communication research. When the work was over, they were released off the southeast Atlantic.

What was learned from the Janus Project was that dolphins could demonstrate knowledge of a whistle language, and could associate whistle words with objects and actions. But, it was not interactive communication. Also, the choice of whistles for the language was not the best one. Most of the dolphin's natural communication occurs as high bandwidth clicks, which are higher in frequency than whistles. But whistles are easier to produce accurately (by humans), than clicks. Dolphins produce both clicks and whistles by shuttling air rapidly back and forth through chambers near their blowhole.

Lilly would sometimes like to listen to the dolphins on the underwater speaker in the isolation tank on the Marine World back lot. He would put a microphone near his head and whistle and click back to the dolphins. Floating in the isolation tank gave insights to a floating intelligence, and led to fantastic swim sessions with Joe and Rosie.

More personally meaningful communication exchanges occurred in the swim interactions than the language sessions. John and Toni Lilly had droves of friends that came through Marine World to swim with the dolphins. People like Ram Dass, Patricia Sun, Joan Halifax, Olivia Newton John, and others swam with Joe and Rosie. Some were delighted by the experience, and some were scarred by it.

Working with dolphins awakened the fantasy of a human/dolphin community, a place on the edge of the ocean where dolphin and human families could grow up together. We saw that dolphins get along well with children when Janet Lederman, director of Esalen Institute, brought the Gazebo kids to swim with Joe and Rosie. The dolphins were careful with children, and with handicapped people. The children love being in the water with dolphins. The dream of living in an ocean community and continuing the interspecies communication work is still alive.

– courtesy Ed Ellsworth
The Dolphin Network
spring 1988

Jennie O'Conner and Toni Lilly swimming at Marine World with Joe and Rosie.

Burgess Meredith, John Lilly, and John Kent (background) in the JANUS control trailer.

Steve Castillo

Burgess Meredith gets an Orca kiss.

Steve Castillo

John Lilly makes contact.

Simulation of
the Future of Man,
Dolphin, and Whale

In solitude John contemplated alternative futures for Man and the Cetacea. His skepticism and pessimism dominated him at the beginning of this consideration.

"Let me assume that we do break the communication barrier by modern technical methods and establish communication between us and them. Can this be done openly, publicly, with the cooperation of the new generation of humans? In my current pessimistic mood, I visualize scenarios in which those in power—the corporations, the military, the covert intelligence services—solve this problem before it is solved publicly. What would happen under those circumstances?"

In this first alternative scenario for the future, one can visualize what will happen on the basis of public knowledge, without knowing what is being done under top security.

The Naval Undersea Center has released two films; one shows human control of dolphins by means of Wayne Batteau's transphonemator. A human is shown operating this device, talking to a dolphin in Hawaiian. The dolphin responds by moving in certain directions and accomplishing a specific task. For some unknown reason the film leaves out the dolphin's vocal responses. When this device was demonstrated to John by Batteau several years ago, the dolphin responded to the commands by whistles which the device turned into human vowel sounds. At that time the dolphin did twenty separate actions in response to the human voice. Batteau had spent two weeks in the Miami laboratory with John before he developed this method, seeing the results of dolphin research there. The effectiveness of the device in training dolphins was shown graphically in the Naval Undersea Center's film.

The second N.U.C. film is entitled "Deep Ops." A pilot whale with special devices retrieves a missile from the ocean floor. The sound track on the film says, "No unusual methods have been used to train this whale." Nothing is said about using the human voice for the control of the whale. Without showing it, the sound track also says that a killer whale was trained in the same way.

On national TV on the *Tomorrow* show, Tom Snyder interviewed a man named Michael Greenwood. Greenwood claimed that he had been involved in Navy-sponsored research in which dolphins were trained to place packages on the bottoms of ships and submarines. These packages could contain either explosives or devices to monitor what was going on in the ship or the submarine, presumably nuclear-powered ones.

In the motion picture *The Day of the Dolphin*, dolphins are shown placing such an explosive package on the bottom of a yacht in which a group of secret agents who stole the dolphins are waiting. The yacht blows up and the dolphins return to their "kindly scientist," played by George C. Scott. According to a brochure which the Avco Company released with the movie, it was suggested that this movie was based on the research of Dr. John Lilly and on a novel (translated from French) written by Robert Merle. Presumably the dolphins here were taught to speak a primitive form of English with open mouths. To those who have heard the real tapes of real dolphins attempting to speak English (with open blow-hole and closed mouth), it was obvious that the movie faked the voice of the dolphin in the air.

The Naval Undersea Center's films imply that the Navy has gone much farther than the films show, released as they are for public education and justification of the Center's budget. The

Greenwood allegations that the Navy and the CIA have been mis-using dolphins in covert operations, as demonstrated in the movie *Day of the Dolphin*, establish a rather pessimistic alternative scenario for the future of the relations between dolphin and Man.

To John this was public evidence of how his early research with dolphins could be used in the service of those who put making war above obtaining new knowledge. He realized the power of desperate men fighting other desperate men. Their adequately funded secret work and the entertainment based on their work created an impression of dolphin research in the public mind which was counter to the evolution of good relations between Man and Cetacea.

One imagines that Man would never really communicate with the Cetacea under the auspices of such agencies. The aims of their research are to use the Cetacea in the service of Man's warfare on Man. In the present human consensus reality, large amounts of money are devoted to this research and engineering.

Such research stems directly from the current biological dogma that Cetacea are something less than Man, that Cetacea can be trained in the service of Man, that Cetacea have no culture of their own, no intelligence, no sentience, no compassion.

The present laws, which include Cetacea, assume that they are "an economic resource for the use of Man." Such laws are epitomized in the United States Marine Mammal Protection Act of 1972 and in the Endangered Species Act of 1973. The International Whaling Commission is founded on this same belief system, i.e., that the other species are for the economic exploitation of Man.

Currently there are several tens of groups of persons who wish "to conserve the species of Earth" including the Cetacea. There is, for instance, a strong antiwhaling movement. Most of these groups are staffed by compassionate human beings who fight against the killing of the Cetacea. Insofar as can be determined, they are not united. From group to group the philosophies vary. None have come straight out in a public display of belief in the high intelligence, the compassion, and the cultural attainments of the Cetacea.

There are a few members of the younger generation who believe in the intelligence and ancient wisdom of the dolphins and whales. These people are very much in the minority.

The current scientific consensus about the dolphins and the whales is generated by zoologists and biologists who cannot believe in, and therefore do not write about, the high intelligence of the Cetacea. Such ideas are relegated to those enthusiastic younger people who have not been trained in the dogma of biological traditions.

There is a fair number of individuals working for the appreciation of the intelligence of the Cetacea. Many of these individuals have communicated with John. Most of them have asked him not to give their names publicly. Given the present human consensus reality and the published scientific dogma, these individuals do not feel that it is yet appropriate to acknowledge the new belief system in public.

If the trends of the past continue, there is no future for the Cetacea. The whaling industry has assured the demise of the older, larger whales who contain the cetacean wisdom of the past and have taught it to the younger Cetacea. In the last century there were sperm whales up to eighty feet in length; currently, none have been seen larger than fifty feet. The obvious conclusion is that the old sperm whales are gone and that with them a good deal of the sperm-whale knowledge and wisdom has been extinguished. One can assume not only that the older whales educate the younger ones but that there is also interspecies communication in the sea among the Cetacea. Man has missed the opportunity to share in this ancient culture; the ancient cultures in the New World were eliminated in a similar fashion during the Spanish conquest of Mexico and South America. The Mayan, Aztec, and Inca civilizations and records were destroyed as heretical and non-Christian.

So the Cetacean culture may have been, or soon will be, destroyed.

John shook himself, attempting to get out of the pessimistic despair into which he was thrown by a review of these facts. Enmeshed in thoughts and feelings about Man's inhumanity to Man,

Man's inhumanity to other species, he found it difficult to conceive of alternatives for the future of the Cetacea.

He thought, "What is the meaning of humanity, of humanitarian? These words are derived from Man's ideals about his own nature. 'Humanitarian' has come to imply high ideals. When applied to Man's relations to the animals this concept assumes that somehow or other Man is superior to the other species. In colleges and universities, the humanities are the studies of Man himself and his literary productions about himself. With such labels Man chases his own tail, his own past. The self-referential nature of Man's studies is excruciatingly obvious."

John thought, "Our task is not to be humanitarians, not so much to participate in conservation of species as to be willing, positive participants in the future evolution of all species, including Man, upon the planet Earth. Man must come down from his throne and realize that his future is coextensive with the future of all species. Instead of functioning as a reluctant caretaker, the zookeeper of Earth, it is time for Man to change his beliefs and become what he is, another species that desires survival not at the expense of but in concert with the other organisms of the planet.

"Alternative futures for Man, then, need an opening of his communication systems, currently devoted exclusively to inter-human problems, to include the other capable species in his communication. Man needs a new humility, a new belief in the abilities of these species to communicate with him. He needs to be freed of his suffering from interspecies deprivation."

John felt his own head, the size of his brain. He realized the small size of the brains of all humans. He visualized the size of the brains of dolphins, the huge brains of the Cetacea. He was aware of his own small mind compared with the huge minds resident in the brains of dolphins and whales.

He thought, "Let me forget the barriers to alternative futures for us. Let me forget pessimism in regard to those futures. Let us construct an alternative future."

Several groups of young humans establish communication

with a few dolphins. At the beginning, these are young dolphins who have not yet learned all of dolphin culture. The young dolphins are confined and then released. New young dolphins are captured, taught, and released. In certain installations the older dolphins begin to approach the human researchers, intrigued by what the younger dolphins have told them about the new breed of Man. The older dolphins try to teach these young humans the dolphin ethics, the dolphin languages.

The whales begin approaching these groups, informed by the dolphins of what is occurring. These exchanges are taking place in remote locations, safe from the whalers, and the aggressive human groups.

Over the next few years these young researchers break through into the multiple languages of the Cetacea. They begin to publish the long sagas, the teaching stories of the dolphins and the whales. They find that these stories go back thirty million years, passed on by the traditions taught by the Cetacea to one another.

These stories include the total interdependence of the organisms of the sea. They reveal the means by which this interdependence was realized, the reasons why in the sea there is respect among species, a respect not present among the species on land. The tolerance of the larger-brained species for the smaller-brained ones is elucidated. The rules of survival as worked out in the sea over a thirty-million-year period are taught to Man.

In these stories are histories of cataclysms of the planet Earth, huge earthquakes, the shifting of continental masses, the destruction of seas; the violence of the surface of flooded continents is recounted. The survival over this extended period and the evolution of the large pelagic brains, despite these cataclysms, is carefully explained. The interconnected oceans allowed sufficient numbers of Cetacea to travel away from the sites of the cataclysms and save themselves to tell these stories to their descendants. Time after time the species on land have been wiped out while those in the oceans were saved. The cushioning effects of vast masses of water assured the long-term survival of the increasingly larger brains and their vehicles.

The whales tell the men stories of their own evolution in the sea. They tell of the demise of the organisms of the land during the cataclysms. They tell Man of past contacts between his species and theirs: in several remote locations for very short periods of time, they have established contact with early Man.

The Cetacea tell stories of the arrival of strange vehicles from outer space which landed in the sea. Many of these burned up coming through the atmosphere; a few landed intact and their appearance and their occupants are described in the whale stories. The whales established communication with a few of these arrivals from outer space before the extraterrestrials left again. Thus the whales have been educated about interplanetary travel by these visitors from beyond Earth.

Realizing that Man is learning how to travel off the planet, the whales describe to Man the extraterrestrial visitors and their vehicles. They teach Man the lessons they have learned from these strange life forms from elsewhere. The whales suggest new directions for research into space travel.

They push the evolution of Man's science beyond its present limits. They furnish Man with hints for means of communication far beyond anything that he has accomplished in his own human-centered science of the past. The whales teach Man the lesson that survival in the galaxy depends on communication of the basic survival programs for all life, irrespective of its origins.

The men of that future day solve their national and international disputes in the light of the new knowledge. They establish communication with far civilizations elsewhere in the galaxy. They are taught appropriate methods of communication with these civilizations by the whales. They learn how to join in the networks of communication throughout the galaxy. They are made aware of the existence of intelligences far greater than that of whale or Man. The men of that future time begin to realize their proper place in the galaxy and then in the universe as a whole.

The new God of these new men becomes large enough to encompass the whole Universe. The new evolutionary God gives them His dictum, "Do not construct lesser gods than I am. Do not worship lesser gods than I."

Coming out of his reverie, John resumed his place as a human of the twentieth century. He thought, "I have imagined a future. Now let us get on with the nitty-gritty of breaking the communication barrier with the Cetacea."

Epilogue

This book has ended. The story of life on this planet as we know it continues. The seeds for the future plants are in the story told in the book. The plants of the future grow. Will they flower into what the author wishes them to be? Only the Earth Coincidence Control Office (ECCO) and its Beings seem to know.

In the interim, after completion of the book, much happened to John and Toni. They became immersed in the creation of the program to test the mutual intelligence of Cetaceans and humans (including themselves). The bare essential beginnings of the human-dolphin tests, under way. Through the help of friends, a computer has been purchased, put into operation, and the necessary software research initiated.

John and Toni. They wondered: Can enough informed, interested, dedicated humans and enough money be found to initiate and to continue the full program of interspecies communication?

As they recount in *The Dyadic Cyclone*, chapter Zero, human agents of ECCO are required to exert their best efforts, their best intelligence, in the service of ECCO; to expect the unexpected every minute, hour, day, week of their lives. ECCO controls long-term coincidence patterns; agents are expected to control short-term ones.

With their friends, John and Toni worked on the short-term coincidence patterns; they made real the dream of communication with dolphins, porpoises, and whales through computer sessions of learning with them.

A new book also seems to be gestating; so life continues, etc., etc., etc., to its future ending.

Malibu, 1977

Dyadic Cyclone

John and Toni Lilly, the dyadic cyclone.

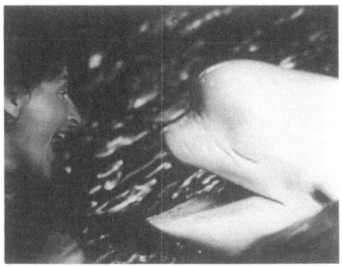

Toni Lilly engages in an animated dialogue with Beluga dolphin.

Faustin Bray

Antonietta L. Lilly
11/25/28 – 1/28/86

When Toni told us about her body's advanced bone cancer condition, she spoke with a light voice, recounting a recent conversation with Richard Ram Dass Alpert during which he relayed what his guru told him: "Death has been given a bad rap, you know. It's like taking off a shoe that's too tight."

In the following one-and-a-half months, Toni distributed that wisdom and pieces of a future script to her friends and relations, saying that she believed in models and we needed one for leaving the body with dignity. She was ready and willing.

She called up a cosmic overlay of an Albanian/Sicilian Dantes's *Inferno*, *Pergatorio*, and *Paradiso* in the operactic crossing-over of a Mediterranean tribal matriarch, mother goddess, sister Diana — huntress with hounds, and Calypso Nymph Mischievious playing in the waves of uncertainty.

We A.L.L. have a part, we are assured, in the archetypal musical chairs, with metamorphoses and alchemical transmutation keeping the natural rhythm. She transferred that sense of . . . "it's all happening and you are here, there, and everywhere." The continuation of the life dance is primary and one may flamboyantly hostess the eternal party on decks festooned with multi-colored streamers fluttering in gentle breezes. Permissionaries of courtly love, questers for the Philosopher's Stone, sway hand in hand with California style-setters, transcending limits at the spring-fed watering hole of conceptualists from all facets of frontline thinking/being.

— Faustin Bray

Front Runner

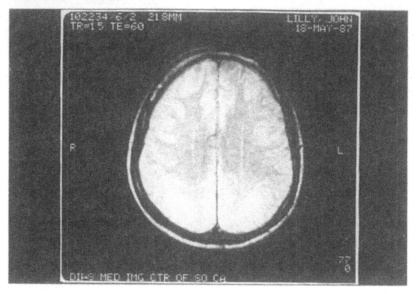

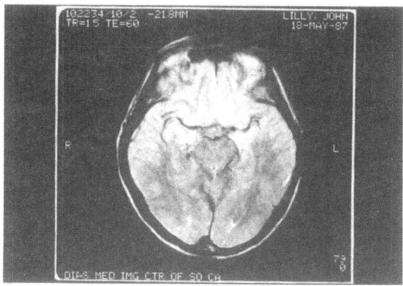

NMR (Nuclear Magnetic Resonance) scan of the brain of John Lilly.

Friends

Many years ago I found out where John Lilly lived and, uninvited, went to see him. He is a very private man, still he received me and we became friends. In fact, almost as a member of his family. I followed the events described in his books *Center of the Cyclone*, *Deep Self*, and, more recently, *The Scientist*. Presently I am trying, as best I can, to be of practical assistance to him through the Human/Dolphin Foundation, which a few of us established three years ago. This foundation is located in a high canyon above the Malibu Hills in southern California. It is near the residence of John Lilly and his wife, Toni.

From time to time Toni and John take off to give a lecture and/or workshop or for a social visit; but mostly, for seven days out of the week, they attend the data that is being accumulated and the blueprints that are being drawn for Project Janus (Interspecies Communication with Cetaceans).

Adjacent to the Lilly house is a small one-and-a-half room laboratory. Here, five days a week, a handful of young scientists, who have come from different parts of the United States and Canada, work, without pay, under Lilly's supervision, or, more simply, to work on the various computers, hydrophones, and calculating paraphernalia that the foundation has purchased (which you have read about in this book), and to prepare the results for scientific testing.

One feels privileged to be in the vicinity of this activity. The word is out that important steps are being taken, and, indeed, important people come and go.

The implications of a possible breakthrough in establishing communication with an alien species whose brain size is larger than our own (whether on this planet or another) are enormous, and the consequences of finally ending what has been called the "long loneliness of man" on this Earth, in our time, would, of course, be epic.

What it amounts to is that we are in a race to speak to the whales and dolphins before they are destroyed. Like a Greek drama the tension is great and the outcome is uncertain.

The late John Steinbeck wrote in *The Sea of Cortez*: "It is a good thing for a man to look down at the tide-pools, then up to the stars, then back to the tide-pools again."

John Lilly has put it another way, and it was this statement that drew many of us to him: "In the province of the mind, what one believes to be true either is true or becomes true within certain limits. These limits are found to be beliefs to be transcended."

Burgess Meredith

   Lisa Lyon Lilly is John Lilly's adopted daughter, ally and associate. A
conceptual artist and classical academic, she is best recognized as a star
athlete and the first World Women's Body Building Champion. Lisa
told us, " He said to me once, 'Forgive me darling if I can't live down to
your expectations.' Only one adjective can be absolutely excluded in
attempting to describe him. Never is he boring."

Nancy Ellison

"As a couple, Toni and John were very intuitive and supportive of their friends. I remember one time when I was feeling down. They arrived unexpectedly just at the right time with just the right bottle of champagne. You can call it serendipity if you like.

John introduced me when Sierra University presented me with an honorary degree. He spoke of my writing about Aldous's death in my book, *This Timeless Moment*. It was so unexpected for an occasion like that. He used such beautiful words. I meant a lot to me."

- Laura Huxley

Faustin Bray

Richard Feynman.

Mr. Lilly introduced me to the tanks. I realized that other people had found the sense-deprivation tank somewhat frightening, but to me it was a pretty interesting invention. Mr. Lilly had a number of different tanks, and we tried a number of different experiments. We also had numerous discussions about hallucinations, the imagination, and what represents true reality. The Lillys were very interesting people; I enjoyed them very, very much.

— courtesy of Richard Feynman
from *Surely You're Joking, Mr. Feynman*

Toni Lilly

Robin Williams, who helped with the JANUS project.

courtesy of the Lilly family

John Lilly, with his brother David Lilly, now Vice President of Finance at University of Minnesota in Minneapolis.

I met John Lilly at Laura Huxley's house — before John met his wife Toni. John and Toni and my family have been close ever since.

I love his book *The Scientist*. Since then, I have had a few ideas regarding my earlier comment about the book.

John is a real scientist and real doctor who believes a doctor shouldn't prescribe a medicine unless they have tried it themself.

John has been involved with experiments with chemicals, in order to find out if there is another reality besides the ordinary one.

John is now involved with experiments with vitamin C.

I believe that certain things cannot [ ] be proven scientifically. Certain things need not be proven scientifically but experientially.

Many scientists, including Einstein, arrive at their (insights) conclusions through intuition.

John Penham, overstates the need for scientific proof.

I am personally aware of other realities, which I have experienced since knowing Lilly.

Benjamin Weininger, M.D.
Lifetime Member of American Psychiatric Association

Faustin Bray

Neurotrekking with Barbara Clarke and friends by the Molniya satellite linkup to the Soviet Union in San Francisco.

Bibliography (*Books by John C. Lilly, et al.*)

1. Borsook, Henry, J. Dubnoff and John C. Lilly. 1941. "The Formation of Glycocyamine in Man and Its Urinary Excretion." *J. Biol. Chem. 138*:405–419
2. Lilly, John C. 1942. "The Electrical Capacitance Diaphragm Manometer." *Rev. Sci. Instrum. 13*:34–37
3. Lilly, John C., and Thomas F. Anderson. 1943. "A Nitrogen Meter for Measuring the Nitrogen Fraction in Respiratory Cases." Nat'l. Research Council, CMR–CAM Report #299 PB 95882 Library of Congress. Photoduplication Service, Publication Board Project, Washington 25, DC.
4. Lilly, John C. 1944. "Peak Inspiratory Velocities During Evercise at Sea Level" in *Handbook of Respiratory Data in Aviation.* Nat'l. Research Council, Wash., DC.
5. Lilly, John C. and Thomas F. Anderson. 1944. "Preliminary Studies on Respiratory Gas Mixing with Nitrogen as a Tracer Gas." *Am. J. Med. Sci. 208*:136
6. Lilly, John C., John R. Pappenheimer and Glenn A. Millikan. 1945. "Respiratory Flow Rates and the Design of Oxygen Equipment." *Am. J. Med. Sci. 210*:810
7. Lilly, John C. 1946. "Studies on the Mixing of Gases Within the Respiratory System with a New Type Nitrogen Meter." (Abstract) Fed. Proc. *5*:64
8. Lilly, John C., Victor Legallais and Ruth Cherry. 1947. "A Variable Capacitor for Measurements of Pressure and Mechanical Displacements: A Theoretical Analysis and Its Experimental Evaluation." *J. Appl. Phys. 18*:613–628

Bibliography

9. Lilly, John C. 1950. "Flow Meter for Recording Respiratory Flow of Human Subjects" in *Methods in Medical Research. Vol.* 2:113–122. J. H. Comroe, Jr., Ed. Year Book Publishers, Inc., Chicago

10. Lilly, John C. 1950. "Physical Methods of Respiratory Gas Analysis" in *Methods of Medical Research. Vol.* 2:131–138. J. H. Comroe, Jr., Ed. Year Book Publishers, Inc., Chicago

10A Lilly, John C. 1950. "A 25-Channel Recorder for Mapping the Electrical Potential Gradients of the Cerebral Cortex: Electro—Iconograms." Electrical Engineering. A.I.E.E., Annual Index to Electrical Engineering 69:68–69

11. Lilly, John C. 1950. "Respiratory System: Methods: Gas Analysis." in *Medical Physics. Vol.* 2:845–855. O. Glasser, Ed. Year Book Publishers, Inc., Chicago

12. Lilly, John C. 1950. "Mixing of Gases Within Respiratory System with a New Type of Nitrogen Meter." *Am. J. Physiol.* 161:342–351

13. Lilly, John C. 1950. "A Method of Recording the Moving Electrical Potential Gradients in the Brain. The 25-Channel Bavatron and Electro-Iconograms." (A.I.E.E.-IRE Conf. on Electronic Instrumentation in Nucleonics and Medicine). Am. Inst. of Electr. Eng., New York. S–33:37–43

14. Lilly, John C. 1950. "Moving Relief Maps of the Electrical Activity of Small (1 cm²) Areas of the Pial Surface of the Cerebral Cortex." *EEG. Clin. Neurophysiol.* 2:358

15. Chambers, William W., George M. Austin, and John C. Lilly. 1950. "Positive Pulse Stimulation of Anterior Sigmoid and Precentral Gyri; Electri Current Threshold Dependence on Anesthesia, Pulse Duration and Repetition Frequency." (Abstract). Fed. Proc. 9:21–22

16. Lilly, John C. and William W. Chambers. 1950. "Electro-Iconograms from the Cerebral Cortex (cats) at the Pial Surface: 'Spontaneous' Activity and Responses to Endorgan Stimuli Under Anesthesia." (Abstract). Fed. Proc. 9:78

17. Lilly, John C. 1950. "Moving Relief Maps of the Electrical Activity of Small (1 cm²) Areas of the Pial Surface of the Cerebral Cortex. Anesthetized Cats and Unanesthetized Monkeys" (Abstract). Proc. 18th Int'l. Physiol. Congress, Copenhagen. P. 340-351

18. Lilly, John C. 1951. "Equipotential Maps of the Posterior Ectosylvian Area and Acoustic I and II of the Cat During Responses and Spontaneous Activity" (Abstract). Fed. Proc. 10:84

19. Lilly, John C. and Ruth Cherry. 1951. "An Analysis of Some Responding and Spontaneous Forms Found in the Electrical Activity of the Cortex." *Am J. Med. Sci.* 222:116–117

20. Lilly, John C., and Ruth Cherry. 1951. "Traveling Waves of Action and of Recovery During Responses and Spontaneous Activity in the Cerebral Cortex." *Am. J. Physiol. 167*:806

21. Lilly, John C. 1952. "Forms and Figures in the Electrical Activity Seen in the Surface of the Cerebral Cortex" in *The Biology of Mental Health and Disease* (1950 Milbank Mem. Fund Symposium). Paul B. Hoeber, Inc., New York. P. 205–219

22. Lilly, John C., George M. Austin, and William W. Chambers. 1952. "Threshold Movements Produced by Excitation of Cerebral Cortex and Efferent Fibers with some Parametric Regions of Rectangular Current Pulses: (Cats and Monkeys)." *J. Neurophysiol. 15*:319–341

23. Lilly, John C. and Ruth Cherry. 1952. "New Criteria for the Division of the Acoustic Cortex into Functional Areas" (Abstract). Fed. Proc. *11*:94

24. Lilly, John C., and Ruth Cherry. 1952. "Criteria for the Parcelation of the Cortical Surface into Functional Areas" (Abstract). *EEG. Clin. Neurophysiol. 4*:385

25. Lilly, John C. 1953. "Significance of Motor Maps of the Sensorimotor Cortex in the Conscious Monkey." (Abstract). Fed. Proc. *12*:87

26. Lilly, John C. 1953. "Discussion of Paper by Lawrence S. Kubie; Some Implications for Psychoanalysis of Modern Concepts of the Organization of the Brain." *Psychoanalytic Q. 22*:21–68

27. Lilly, John C. 1953. Review of book by W. Ross Ashby: *Design for a Brain.* John Wiley and Sons, Inc., New York. *Rev. of Sci. Instrum. 24*:313

28. Lilly, John C. 1953. "Functional Criteria for the Parcelation of the Cerebral Cortex." Abstracts of Communications, XIX Int'l. Physiol. Cong., Montreal, Canada. P. 564

29. Lilly, John C. 1953. Recent Developments in EEG Techniques: Discussion. (Third Int'l. EEG Cong. 1953. Symposia). EEG Clin. Neurophysiol. Suppl. *4*:38–40

30. Lilly, John C. 1954. Critical Discussion of Research Project and Results at Conference in June 1952 by Robert G. Heath and Research Group at Tulane Univ. in Robert G. Heath, et al. *"Studies in Schizophrenia: A Multidisciplinary Approach to Mind–Brain Relationships."* P. 528–532

31. Lilly, John C. 1954. "Instantaneous Relations Between the Activities of Closely Spaced Zones on the Cerebral Cortex: Electrical Figures During Responses and Spontaneous Activity." *Am. J. Physiol. 176*:493–504

32. Lilly, John C., and Ruth Cherry. 1954. "Surface Movements of Click Responses from Acoustic Cerebral Cortex of Cat: Leading

and Trailing Edges of a Response Figure." *J. Neurophysiol.* 17:521–532

33. Lilly, John C. 1954. Discussion, Symposium on Depth Electrical Recordings in Human Patients. Am. EEG Soc. Neurophysiol. 6:703–704

34. Lilly, John C., and Ruth Cherry. 1955. "Surface Movements of Figures in Spontaneous Activity of Anesthetized Cerebral Cortex: Leading and Trailing Edges. *J. Neurophysiol.* 18:18–32

35. Lilly, John C., John R. Hughes, and Ellsworth C. Alvord, Jr., and Thelma W. Galkin. 1955. Brief. "Noninjurious Electric Waveform for Stimulation of the Brain." *Science* 121:468–469

36. Lilly, John C., John R. Hughes, and Ellsworth C. Alvord, Jr., and Thelma W. Galkin. 1955. "Motor Responses from Electrical Stimulation of Sensorimotor Cortex in Unanesthetized Monkey with a Brief, Noninjurious Waveform" (Abstract). Fed. Proc. 14:93

37. Lilly, John C. 1955. "An Anxiety Dream of an 8-Year-Old Boy and Its Resolution." *Bul. Phila. Assn. for Psychoanal.* 5:1–4

38. Lilly, John C. 1955. Review of book by Robert G. Heath, et al., 1954. *Studies in Schizophrenia: A Multidisciplinary Approach to Mind–Brain Relationships.* Harvard Univ. Press. EEG Clin. Neurophysiol. 7:323–324

39. Lilly, John C., John R. Hughes, Thelma W. Galkin and Ellsworth C. Alvord, Jr. 1955. "Production and Avoidance of Injury to Brain Tissue by Electrical Current at Threshold Values." *EEG Clin. Neurophysiol.* 7:458

40. Lilly, John C. 1956. "Effects of Physical Restraint and of Reduction of Ordinary Levels of Physical Stimuli on Intact Healthy Persons." 13–20 & 44, in *Illustrative Strategies for Research on Psychopathology in Mental Health, Symposium No. 2.* Group for the Advancement of Psychiatry. New York. P. 47

41. Lilly, John C., John R. Hughes, and Thelma W. Galkin. 1956. "Gradients of Motor Function in the Whole Cerebral Cortex of the Unanesthetized Monkey" (Abstract). Fed. Proc. 15

42. Lilly, John C., John R. Hughes, and Thelma W. Galkin. 1956. "Physiological Properties of Cerebral Cortical Motor Systems of Unanesthetized Monkey" (Abstract). Fed. Proc. 15

43. Lilly, John C. 1956. "Mental Effects of Reduction of Ordinary Levels of Physical Stimuli on Intact. Healthy Persons" in *Psychiat. Res. Reports 5.* American Psychiatric Assn., Wash., DC. P. 1–9

44. Lilly, John C., John R. Hughes, and Thelma W. Galkin. 1956. "Some Evidence of Gradients of Motor Function in the Whole Cerebral Cortex of the Unanesthetized Monkey" (Abstract). *Proc. 20th Int'l. Physiol. Congress.* P. 567–568

45. Lilly, John C. 1956. "Distribution of 'Motor' Functions in the Cerebral Cortex in the Conscious, Intact Monkey." *Science.* *124*:937

46. Lilly, John C. 1957. "Some Thoughts on Brain–Mind and on Restraint and Isolation of Mentally Healthy Subjects. (Comments on Biological Roots of Psychiatry by Clemens F. Benda, M.D.)" *J. Phila. Psychiatric Hosp.* 2:16–20

47. Lilly, John C. 1957. "True Primary Emotional State of Anxiety—Terror—Panic in Contrast to a 'Sham' Emotion or 'Pseudo—Affective' State Evoked by Stimulation of the Hypothalamus" (Abstract). Fed. Proc. *16*:81

48. Lilly, John C. 1957. "Learning Elicited by Electrical Stimulation of Subcortical Regions in the Unanesthetized Monkey." *Science.* *125*:748

49. Lilly, John C. 1957. Review of book by Donald A. Scholl. 1956. *The Organization of the Cerebral Cortex.* Methuen and Co., Ltd., London and John Wiley and Sons, Inc., New York. *Science.* *125*:1205

50. Lilly, John C. 1957. "A State Resembling 'Fear–Terror–Panic' Evoked by Stimulation of a Zone in the Hypothalamus of the Unanesthetized Monkey." *Excerpta Medica.* Special Issue, Abstracts of Fourth Int'l. Cong. EEG and Clin. Neurophysiol. and 8th Meeting of the Int'l. League Against Epilepsy. Brussels. P. 161

51. Lilly, John C. 1957. " 'Stop' and 'Start' Systems" *in Neuropharmacology.* Transactions of the Fourth Conference, Josiah Macy, Jr., Foundation, Princeton, N.J. (L.C. 55–9013). P. 153–179

52. Lilly, John C. 1958. "Learning Motivated by Subcortical Stimulation: The 'Start' and 'Stop' Patterns of Behavior." 705–721. *Reticular Formation of the Brain.* H. H. Jasper, et al. Eds. Little, Brown and Co., Boston. P. 766

53. Lilly, John C. 1958. "Correlations Between Neurophysiological Activity in the Cortex and Short-Term Behavior in the Monkey," in *Biological and Biochemical Bases of Behavior* (Univ. of Wis. Symposium. 1055) H. F. Harlow and C. N. Woolsey, Ed. Univ. of Wis. Press, Madison, Wis. P. 83–100

54. Lilly, John C. 1958. "Development of a Double–Table–Chair Method of Restraining Monkeys for Physiological and Psychological Research." *J. Appl. Physiol.* 12:134–136

55. Lilly, John C. 1958. "Simple Percutaneous Method for Implantation of Electrodes and/or Cannulae in the Brain." (Abstract.) Fed. Proc. *17*:97

56. Lilly, John C. 1958. "Electrode and Cannulae Implantation in the Brain by a Simple Percutaneous Method." *Science.* *127*:1181–1182

Bibliography

57. Lilly, John C. 1958. "Some Considerations Regarding Basic Mechanisms of Positive and Negative Types of Motivations." *Am. J. Psychiat.* 115:498–504

58. Lilly, John C. 1958. "Rewarding and Punishing Systems in the Brain" in *The Central Nervous System and Behavior.* Transactions of the First Conference, Josiah Macy, Jr., Foundation, Princeton, N.J. (L.C. 59–5052.) P. 247–303

59. Lilly, John C. 1959. " 'Stop' and 'Start' Effects in *The Central Nervous System and Behavior.* Transactions of the Second Conference, Josiah Macy, Jr., Foundation and National Science Foundation, Princeton, N.J. (L.C. 59–5052.) P. 56–112

60. Lilly, John C. 1960. "Learning Motivated by Subcortical Stimulation: The 'Start' and The 'Stop' Patterns of Behavior. Injury and Excitation of the Brain by Electrical Currents." Chapter 4 in *Electrical Studies on the Unanesthetized Brain.* E. R. Ramsey and D. S. O'Doherty, Eds, Paul B. Hoeber, Inc., New York. P. 78–105

61. Lilly, John C. 1960. Contributing Discussant—The Central Nervous System and Behavior. Transactions of the Third Conference Josiah Macy, Jr., Foundation, Princeton, N.J. (L.C. 59–5052.)

62. Lilly, John C. 1960. "The Psychophysiological Basis for Two Kinds of Instincts." *J. Am. Psychoanalyt. Assoc.* Vol. 8: P. 659–670

63. Lilly, John C. 1960. "Large Brains and Communication." Paper Presented to the Philadelphia Assoc. for Psychoanalysis.

64. Lilly, John C. 1961. "Injury and Excitation by Electric Currents." Chapter 6 in *Electrical Stimulation of the Brain.* Daniel E. Sheer, Ed., Univ. of Texas Press for Hogg Foundation for Mental Health, Austin, Texas. P. 60–64

65. Lilly, John C. and Jay T. Shurley. 1961. "Experiments in Solitude, in Maximum Achievable Physical Isolation with Water Suspension of Intact Healthy Persons." (Symposium, USAF Aerospace Medical Center, San Antonio, Texas, 1960.) in *Psychophysiological Aspects of Space Flight.* Columbia Univ. Press, New York. P. 238–247

66. Lilly, John C., and Alice M. Miller. 1961. "Sounds Emitted by the Bottlenose Dolphin." *Science.* Vol. 133, P. 1689–1693

67. Lilly, John C., and Alice M. Miller. 1961. "Vocal Exchanges Between Dolphins." *Science.* Vol. 134: P. 1873–1876

68. Lilly, John C. 1961. "Problems of Physiological Research on the Dolphin, *Tursiops*" (Abstract). Fed. Proc. 20:1

69. Lilly, John C. 1961. "The Biological Versus Psychoanalytic Dichotomy." *Bul. of The Phila Assoc. for Psychoanal.* Vol. 11: P. 116–119

70. Lilly, John C. 1961. *Man and Dolphin.* Doubleday & Co., Inc., New York. (L.C. 61–9628)

  1962. *L'Homme et le Dauphin.* Stock Edition, *l'Imprimerie des Dernières Nouvelles de Strasbourg,* Stock, Paris

  1962. *Manniskan och Definen.* Wahlstrom & Widstrand, Bakforlag, Stockholm, Sweden

  1962. *Man and Dolphin.* Victor Gollancz, Ltd., London, England

  1962. *Man and Dolphin.* (The Worlds of Science Series, Zoology.) Pocket Edition, Pyramid Publications, New York

  1963. *Mensen Dolfijn.* Contact—Amsterdam-Druk: Tulp-Zwolle

  1963. *Menneskat og Delfinen.* Nasjonalforlaget, Oslo, Norway

  1965. *Man and Dolphin.* Gakken Books Science Series, Charles E. Tuttle Co., Inc., Tokyo

  1965. *Man and Dolphin.* Izdatelsstvo Mir Zubosky Square 21, Moscow, U.S.S.R.

  1966. *Člŏvek Delfin.* Miroslav Hrncer Vratislav Mazak

  1967. *Man and Dolphin.* Sophia, Bulgaria

71. Lilly, John C. 1962. The Effect of Sensory Deprivation on Consciousness. *Man's Dependence on the Earthly Atmosphere,* Karl E. Schaefer, Ed. MacMillan Co., New York. (L.C. 61–9079.) P. 93–95. (Proceedings 1st Int'l Symp. on Submarine and Space Medicine, New London, Conn., 1958)

72. Lilly, John C., and Alice M. Miller. 1962. "Operant Conditioning of the Bottlenose Dolphin with Electrical Stimulation of the Brain." *J. Comp. & Physiol. Psychol.* Vol. 55: P. 73–79

73. Lilly, John C. 1962. Cerebral Dominance in *Interhemispheric Relations and Cerebal Dominance.* Vernon Mountcastle, M.D., Ed. Johns Hopkins Press, Inc. Baltimore, Md. P. 112–114

74. Lilly, John C., and Alice M. Miller. 1962. Production of Humanoid Sounds by the Bottlenose Dolphin. (Unpublished manuscript.)

75. Lilly, John C. 1962. A New Laboratory for Research on Delphinids. *Assoc. of Southeastern Biologists Bul.* Vol. 9, P. 3–4

76. Lilly, John C. 1962. "Interspecies Communication" in *Yearbook of Science and Technology.* McGraw-Hill. New York. P. 279–281

77. Lilly, John C. 1962. "The 'Talking' Dolphins" in *The Book of Knowledge Annual.* Society of Canada Limited, Grolier, Inc. (This article was updated in the 1969 Yearbook covering the year 1968, pp. 8–15.)

Bibliography

78. Lilly, John C. 1962. "Vocal Behavior of the Bottlenose Dolphin."
    Proc. Am. Philos. Soc. Vol. 106. P. 520–529
79. Lilly, John C. 1962. "Consideration of the Relation of Brain Size
    to Capability for Language Activity as Illustrated by Homo
    sapiens and Tursiops truncatus (Bottlenose Dolphin)." Elec-
    troenceph. Clin. Neurophysiol. 14, no. 3: 424
80. Lilly, John C. 1962. Sensory World Within and Man and
    Dolphin. (Lecture to the Laity, New York Acad. of Med., 1962.)
    Scientific Report no. CRI–0162
81. Lilly, John C. 1963. "Critical Brain Size and Language." Per-
    spectives in Biol. & Med. Vol. 6. P. 246–255
82. Lilly, John C. 1963. "Distress Call of the Bottlenose Dolphin:
    Stimuli and Evoked Behavioral Responses." Science. Vol. 139.
    P. 116–118
83. Lilly, John C. 1963. "Productive and Creative Research with
    Man and Dolphin." (Fifth Annual Lasker Lecture, Michael
    Reese Hospital and Medical Center, Chicago, Ill., 1962). Arch.
    Gen. Psychiatry. Vol. 8. P. 111–116
84. Lilly, John C., and Ashley Montagu. 1963. Modern Whales,
    Dolphins and Porpoises, as Challenges to Our Intelligence in
    The Dolphin in History by Ashley Montagu and John C. Lilly.
    A Symposium given at the William Andrews Clark Memorial
    Library, Univ. of Calif., Los Angeles, Calif. P. 31–54
85. Lilly, John C. 1964. "Animals in Aquatic Environment. Adapta-
    tion of Mammals to the Ocean" in Handbook of Physiology.
    Environment I, Am. Physiol. Soc., Wash., D.C. P. 741–757
86. Jacobs, Myron S., Peter J. Morgane, John C. Lilly and Bruce
    Campbell. 1964. "Analysis of Cranial Nerves in the Dolphin."
    Anatomical Record Vol. 148. P. 379
87. Lilly, John C. 1964. "Airborne Sonic Emissions of Tursiops
    truncatus (M)" (Abstract) J. Acoustical Soc. of Amer. Vol. 36.
    P. 5, 1007
88. Lilly, John C. 1965. "Report on Experiments with the Bottlenose
    Dolphin." (Abstract) Proc. of the Int'l. Symp. on Comparative
    Medicine, Eaton Laboratories, Norwich, Conn. P. 240
90. Lilly, John C. 1965. "Vocal Mimicry in Tursiops. Ability to
    Match Numbers and Duration of Human Vocal Bursts."
    Science Vol. 147 (3655). P. 300–301
91. Lilly, John C. 1966. "Sexual Behavior of the Bottlenose Dolphin
    in Brain and Behavior. The Brain and Gonadal Function." Vol.
    III. R. A. Gorski and R. E. Whalens, Eds., UCLA Forum Med.
    Sci., Univ. of Calif. Press, Los Angeles, Calif. P. 72–76
92. Lilly, John C. 1966. "Sonic-Ultrasonic Emissions of the Bottle-
    nose Dolphin in Whales, Dolphins and Porpoises." Kenneth S.
    Norris, Ed. Proc., 1st Int'l Symp. on Cetacean Research, Wash.,
    DC. 1963. Univ. of Calif. Press. P. 503–509

93. Lilly, John C. 1966. "The Need for an Adequate Model of the Human End of the Interspecies Communication Program." IEEE Military Electronics Conference (MIL-E-CON 9), on Communication with Extraterrestrial Intelligence, Wash., DC. 1965. *IEEE Spectrum 3*, no. 3: P. 159–160

94. Lilly, John C. 1966. Contributing Discussant. Proc. of Conf. on Behavioral Studies. Contractors Meeting, U.S. Army Edgewood Arsenal, Md. 1965. Dept. of the Army EARL Report

95. Lilly, John C. 1966. "Research with the Bottlenose Dolphin" in *Conference on the Behavioral Sciences*, Proc. of Conf. on Behavioral Studies (Contractors Meeting, U.S. Army Edgewood Arsenal, Md. 1965). Dept. of the Army EARL Report

96. Lilly, John C., and Henry M. Truby. 1966. "Measures of Human-*Tursiops* Sonic Interactions" (Abstract). *J. Acous. Soc. of Amer.* Vol. *40,* issue 5. P. 1241

97. Lilly, John C. 1966. "Sound Production in *Tursiops truncatus* (Bottlenose Dolphin)." Conference on Sound Production in Man: Section on Phonation: Control and Speech Communication, New York Acad. of Sciences. Annals of the New York Academy of Sciences. 1968

98. Lilly, John C. 1966. "Intracerebral Reward and Punishment: Implications for Psychopharmacology." Fifth Annual Meeting of American College of Neuropsychopharmacology. San Juan, Puerto Rico. 1968

99. Lilly, John C. 1967. Dolphin-Human Relationship and LSD-25 in *The Use of LSD in Psychotherapy and Alcoholism*. Harold Abramson, Ed. Second International Conference on the Use of LSD in Psychotherapy, South Oaks Research Foundation, Amityville, L.I. 1965. The Bobbs-Merrill Co., Inc., New York. P. 47–52

100. Lilly, John C. 1967. Dolphin's Mimicry as a Unique Ability and a Step Towards Understanding in *Research in Verbal Behavior and Some Neurophysiological Implications*. Kurt Salzinger and Suzanne Salzinger, Eds. Conference on Verbal Behavior, N.Y.C. 1965. Academic Press, New York City. P. 21–27

101. Lilly, John C. 1967. Dolphin Vocalization in Proc. Conf. on *Brain Mechanisms Underlying Speech and Language*. F. L. Darley, Ed. A Symposium held at Princeton, N.J. 1965. Grune and Stratton, New York City. P. 13–20

102. Lilly, John C. 1967. Basic Problems in Education for Responsibility Caused by LSD-25. Proc. of 17th Conf. on Science, Philosophy and Religion in their Relation to the Democratic Way of Life. Clarence H. Fause, Ed. Paper presented in section on Character Education of Scientists, Engineers and Practitioners in Medicine Psychiatry and Science with Strategies for Change. Loyola Univ., Chicago, Ill. 1966

Bibliography

103. Lilly, John C. 1967. *The Mind of the Dolphin.* Doubleday & Co., Inc., New York.
104. Lilly, John C. 1967. "Intracephalic Sound Production in *Tursiops truncatus:* Bilateral Sources" (Abstract). Fed. Proc. 25, no. 2.
105. Lilly, John C. 1967. *The Human Biocomputer: Programming and Metaprogramming.* Miami. Communications Research Institute. 1967. Scientific Report no. CRI 0167.
106. Lilly, John C. 1968. *Programming and Metaprogramming in the Human Biocomputer: Theory and Experiments.* Miami. Communications Research Institute. Scientific Report no. CRI 0167. 2nd Edition
107. Lilly, John C., Alice M. Miller, and Henry M. Truby. 1968. Reprogramming of the Sonic Output of the Dolphin: Sonic-Burst Count Matching. Miami. Communications Research Institute. Scientific Report no. CRI 0267. *J. Acous. Soc. of Amer.* (See number 112.)
108. Lilly, John C., Alice M. Miller, and Henry M. Truby. 1968. "Perception of Repeated Speech: Evocation and Programming of Alternate Words and Sentences." Scientific Report no. CRI 1067
109. Lilly, John C., Alice M. Miller, and Frank Grissman. 1968. "Underwater Sound Production of the Dolphin Stereo-Voicing and Double Voicing." Miami. Communications Research Institute. Scientific Report no. CRI 0367
110. Truby, Henry M., and John C. Lilly. 1967. "Psychoacoustic Implications of Interspecies Communication." Miami. Communications Research Institute. *J. Acous. Soc. of Amer.* Vol. 42: P. 1181. S3 (Abstract.)
111. Lilly, John C., Henry M. Truby, Alice M. Miller, and Frank Grissman. 1967. "Acoustic Implications of Interspecies Communication." Miami. Communications Research Institute. *J. Acous. Soc. of Amer.* Vol. 42: P. 1164. I10 (Abstract.)
112. Lilly, John C., Alice M. Miller, and Henry M. Truby. 1968. "Reprogramming of the Sonic Output of the Dolphin: Sonic Burst Count Matching." *Jnl. of the Acoustical Society of America.* Vol. 43. No. 6. P. 1412–1424
113. Lilly, John C. 1967, 1968, 1972. *Programming and Metaprogramming in the Human Biocomputer, Theory and Experiments.* The Julian Press, Inc. New York.
114. Lilly, John C. 1974. *Programming and Metaprogramming in the Human Biocomputer, Theory and Experiments.* Bantam Books, Inc. New York.
115. Lilly, John C. 1974. *The Human Biocomputer, Theory and Experiments.* Abacus edition, Sphere Books Ltd. London.

116. Lilly, John C. 1972. *The Center of the Cyclone. An Autobiography of Inner Space.* The Julian Press, Inc. New York.
117. Lilly, John C. 1974. *Het Centrum van de Cycloon, Een Autobiografie van de Innerlijke Ruimte.* Wetenschappelijke Uitgeverij b.v. Amsterdam.
118. Lilly, John C. 1972. *The Center of the Cyclone. An Autobiography of Inner Space* (paperback). Bantam Books, Inc. New York.
119. Lilly, John C. 1973. *The Centre of the Cyclone: an Autobiography of Inner Space,* Calder & Boyer, London.
120. Lilly, John C. 1973. *The Centre of the Cyclone: An Autobiography of Inner Space* (paperback). Paladin, Granada Publishing Limited, London.
121. Lilly, John C. 1975. *Simulations of God: The Science of Belief.* Simon and Schuster, New York
122. Lilly, John C. 1975. *Lilly on Dolphins.* Anchor/Doubleday, New York
123. Lilly, John C. and Antonietta L. 1975. *The Dyadic Cyclone.* Simon and Schuster, New York
124. Lilly, John C. The Deep Self: Explorations in Tank-Isolation. (In preparation.)
125. Lilly, John C. Off Center and Return. (In preparation.)

Biographical Data (*Curriculum Vitae*)

John C. Lilly, B.Sc., M.D.

Parents: Richard C. Lilly, Rachel C. Lilly
Birth: 0707 hours, 6 January 1915, St. Paul, Minnesota
Elementary Schools: Grades kindergarten through 4th: Irving Public School; Grades 5, 6, 7: St. Luke's School (Catholic)
Preparatory school: (Forms 2 through 6), St. Paul Academy, St. Paul, Minnesota. Graduation 1933
College: California Institute of Technology, Pasadena, California; (scholarship). Graduation 1938
Medical Schools: Dartmouth Medical School, Hanover, New Hampshire (1938–1940); School of Medicine, University of Pennsylvania, Philadelphia, Pennsylvania. Graduation 1942
University of Pennsylvania: Department of Biophysics and Medical Physics (Eldridge Reeves Johnson Foundation), University of Pennsylvania School of Medicine (1942–1956). Fellow, Associate, Associate Professor (of Medical Physics and of Experimental Neurology)
Psychoanalytic Training: Research Trainee (1949–1957): Training Analyst: Dr. Robert Waelder of the Institute of the Philadelphia Association for Psychoanalysis and the Philadelphia Psychoanalytic Society; the Washington-Baltimore Psychoanalytic Institute: Dr.

Jenny Waelder-Hall, Dr. Lewis Hill; Control Analyst: Dr. Amanda Stoughton

GOVERNMENT SERVICE: Senior Surgeon Grade, United States Public Health Service Commissioned Officers Corps. (1953–1958)

NATIONAL INSTITUTES OF HEALTH, RESEARCH: Section Chief, Section on Cortical Integration in the National Institute of Neurological Diseases and Blindness and in the National Institute of Mental Health, Bethesda, Maryland (1953–1958)

COMMUNICATIONS RESEARCH INSTITUTE: Founder and Director, Saint Thomas, United States Virgin Islands, and Miami, Florida (1959–1968)

NATIONAL INSTITUTE OF MENTAL HEALTH: Research Career Award Fellow (1962–1967)

MARYLAND PSYCHIATRIC RESEARCH CENTER: Catonsville, Maryland. Chief of Psychological Isolation and Psychedelic Research (1968–1969)

ESALEN INSTITUTE: Big Sur, California. Group Leader and Associate in Residence, (1969–1971)

CENTER FOR THE ADVANCED STUDY OF BEHAVIOR: Palo Alto, California. Fellow (1969–1970)

INSTITUTO DE GNOSOLOGIA, Arica, Chile, student (1970–1971)

HUMAN SOFTWARE, INC., Malibu, California. Treasurer (1973–present)

SCIENTIFIC SOCIETIES:

The American Physiological Society (1945–1967)

The American Electroencephalographic Society (1947–1967)

The Institute of Electronic and Electrical Engineers (1951–1967)

The Society of the Sigma Xi (1952–life)

Aerospace Medical Association (1945)

Biophysical Society (charter member to 1967)

The American Medical Authors, Fellow (8 April 1964)

The New York Academy of Science (1949) Fellow (1959)

The Philadelphia Association for Psychoanalysis, Affiliate Member (1958)

California Institute of Technology, Alumni Association (life member)

Order of the Dolphin (1961)

For other society memberships, see listings in *Who's Who* (Marquis) and *American Men of Science*

AWARDS:

California Institute of Technology: Scholar (1933–1935)

University of Pennsylvania School of Medicine: Clark Research Medal (1941)

Biographical Data

Who's Who in the South and Southwest: Citation for outstanding contributions in Science: 16 April 1963

The American Medical Authors, Distinguished Service Award (26 September 1964)

Career Award, National Institute of Mental Health (1962–1967)